AN UPDATE ON POWER QUALITY

Edited by **Dylan Dah-Chuan Lu**

INTECHOPEN.COM

An Update on Power Quality

http://dx.doi.org/10.5772/51135
Edited by Dylan Dah-Chuan Lu

Contributors

Hadeed Ahmed Sher, Khaled E. Addoweesh, Yasin Khan, Sharad W. Mohod, Mohan V. Aware, Mohamed Zellagui, Abdelaziz Chaghi, H.C. Leung, Dylan D.C. Lu, Ahad Mokhtarpour, Heidarali Shayanfar, Mitra Sarhangzadeh, Seiied Mohammad Taghi Bathaee, Raju Ahmed, Mohammad Jahangir Alam

CBS, Edition **2017**

Published by InTech

Janeza Trdine 9, 51000 Rijeka, Croatia

Publishing Process Manager Sandra Bakic

Technical Editor InTech DTP team

Cover Designer InTech Design Team

Additional hard copies can be obtained from orders@intechopen.com

Additional hard copies can be obtained from orders@intechopen.com
An Update on Power Quality, Edited by Dylan Dah-Chuan Lu
p. cm.
ISBN 978-953-51-1079-8

INTECH

open science | open minds

Contents

Preface

There is an upward trend that human activities involve more electric power nowadays and for years to come. And we will soon be facing shortage of electricity supply if we rely solely on non-renewable power generation from fossil fuels such as coals. Renewable energy generation has proven effective in meeting the demand. At the same time it has brought about a few issues to existing electricity infrastructure such as complex power flow due to distributed generations and unstable grid voltage and/or frequency profiles due to local generation and intermittent nature of renewable energy sources such as solar photovoltaic and wind power. Also the increasing penetration of power electronics converters associated with the adoption of renewable energies into the electricity networks has improved the functionality and flexibility in terms of control but meanwhile due to their switching nature the harmonics issue has to be dealt with. It is therefore important to investigate the causes of these power quality issues and explore feasible and cost-effective solutions to assist with the development of present and future electricity networks.

In the following chapters the reader will be introduced to power quality issues and solutions at different sections of the electricity networks, namely, power generation, transmission, distribution and end user stages. The book is divided into three sections: Power Quality Issues and Standards in Electricity Networks; Power Quality Improvements in Transmission and Distribution Systems; and Power Quality Improvement in End Users Stage. A brief discussion of each chapter is as follows.

Chapter 1 provides an overview of the causes, impacts, standards and solutions to voltage and current harmonics problems which are one of the key issues related to power quality. Chapter 2 investigates the challenges of increasing wind generation to the network. The chapter has identified and analyzed briefly several key issues including voltage variations, flickers, switching operation of wind generation, current harmonics and locations of wind turbine. Several solutions have been introduced such as low-voltage ride through capability and international standards. Lastly, but not the least, it describes the purposes of grid code to deal with power quality issues with renewable energy generation.

Flexible Alternating Current Transmission System (FACTS) controllers have been used in transmission and distribution systems to deal with power quality issues. Chapter 3 identifies the limitations in the transmission systems, i.e., angular stability, voltage magnitude, thermal limits, transient stability, and dynamic stability and presents a comparative study of three different series FACTS for power quality compensation of a single transmission line in Eastern Algerian transmissions networks. Chapter 4 applies particle swarm optimization technique to designing a shunt static var compensator (SVC) for a radial distribution feeder. The objective function of SVC placement is to reduce the power loss and keep bus voltages and total harmonic distortion within prescribed limits with minimum cost.

Chapter 5 proposes an electromechanical active power filter which uses a synchronous generator together with a unity power quality conditioner (UPQC). The idea originates from the fact that embedded generation is becoming a viable option in distributed generation and we should utilize this generator as a part of the power quality compensation solution. An algorithm of reference generation has been proposed to work with the two devices and simulation results are reported to verify the effectiveness of the approach. Chapter 6 continues with the previous study but the focus is on the speed to generate the references for UPQC. The general idea is that it uses an adaptive approach to designing the window's width for steady state and transient state operations and to adjusting the reference points for reactive power compensation. The proposed methods reduce the settling time to 1/12 of a cycle and have been verified under voltage sag, swell and load change conditions through MATLAB simulations.

In order to maintain the regulation of AC grid voltage, Chapter 7 investigates different types of AC regulators which include, Solid-state tap changer and steeples control by variac, Solid-tap changer using anti-parallel SCRs, voltage regulation using servo system, phase controlled AC voltage regulator, ferro-resonant AC voltage regulator and switch mode AC voltage regulator. A detailed design of a switch mode AC-AC voltage regulator is presented. Simulation results are reported to show the responsiveness and high power factor of the proposed method.

I hope this book will have been to you an enjoyable reading and a timely update of recent research and development in the field of Power Quality.

Lastly, I would like to thank all the researchers for their excellent works and studies in the different areas of Power Quality.

Dylan Dah-Chuan Lu
University of Sydney
Australia

Section 1

Power Quality Issues and Standards in the Electricity Networks

Chapter 1

Harmonics Generation, Propagation and Purging Techniques in Non-Linear Loads

Hadeed Ahmed Sher, Khaled E. Addoweesh and Yasin Khan

Additional information is available at the end of the chapter

http://dx.doi.org/10.5772/53422

1. Introduction

Industrial revolution has transformed the whole life with advanced technological improvements. The major contribution in the industrial revolution is due to the availability of electrical power that is distributed through electrical utilities around the world. The concept of power quality in this context is emerging as a "Basic Right" of user for safety as well as for uninterrupted working of their equipment. The electricity users whether domestic or industrial, need power, free from glitches, distortions, flicker, noise and outages. The utility desires that the users use good quality equipment so that they do not produce power quality threats for the system. The use of power electronic based devices in this industrial world has saved bounties in term of fuel and power savings, but on the other hand has created problems due to the generation of harmonics. Both commercial and domestic users use the devices with power electronics based switching that draw harmonic current. This current is a dominant factor in producing the harmonically polluted voltages. The "Basic Right" of the user is to have a clean power supply, whereas the demand of utility is to have good quality instrument/equipment. This makes power quality a point of common interest for both the users as well as the utility. Harmonics being a hot topic within power quality domain has been an area of discussion since decades and several design standards have been devised and published by various international organizations and institutions for maintaining a harmonically free power supply. In a wider scenario, the harmonically free environment means that the harmonics generated by the devices and its presence in the system is confined in the allowable limits so that they do not cause any damage to the power system components including the transformers, insulators, switch-gears etc. The deregulation of power systems is forcing the utilities to purge the harmonics at the very end of their generation before it comes to the main streamline and becomes a possible cause of system un-stability. The possible three stage scheme for harmonics control is

- Identification of harmonics sources
- Measurement of harmonics level
- Possible purging techniques

To follow the above scheme the power utilities have R&D sections that are involved in continuous research to keep the harmonics levels within the allowed limits. Power frequency harmonics problems that have been a constant area of research are:

- Power factor correction in harmonically polluted environment
- Failure of insulation co-ordination system
- Waveform distortion
- De-rating of transformer, cables, switch-gears and power factor correction capacitors

The above mentioned research challenges are coped with the help of regulatory bodies that are focused much on designing and implementing the standards for harmonics control. Engineering consortiums like IEEE, IET, and IEC have designed standards that describe the allowable limits for harmonics. The estimation, measurement, analysis and purging techniques of harmonics are an important stress area that needs a firm grip of power quality engineers. Nowadays, apart from the traditional methods like Y-Δ connection for 3rd harmonic suppression, modern methods based on artificial intelligence techniques aids the utility engineers to suppress and purge the harmonics in a better fashion. The modern approaches include:

- Fuzzy logic based active harmonics filters
- Wavelet techniques for analysis of waveforms
- Sophisticated PWM techniques for switching of power electronics switches

The focus of this chapter is to explain all the possible sources of harmonics generation, identification of harmonics, their measurement level as well as their purging/suppression techniques. This chapter will be helpful to all electrical engineers in general and the utility engineers in particular.

2. What are harmonics?

In electrical power engineering the term harmonics refers to a sinusoidal waveform that is a multiple of the frequency of system. Therefore, the frequency which is three times the fundamental is known as third harmonics; five times the fundamental is fifth harmonic; and so on. The harmonics of a system can be defined generally using the eq. 1

$$f_h = hf_{ac} \quad (1)$$

Where f_h is the hth harmonic and f_{ac} is the fundamental frequency of system.

Harmonics follow an inverse law in the sense that greater the harmonic level of a particular harmonic frequency, the lower is its amplitude as shown in Fig.1. Therefore, usually in power line harmonics higher order harmonics are not given much importance. The vital and the most troublesome harmonics are thus 3rd, 5th, 7th, 9th, 11th and 13th. The general expression of harmonics waveforms is given in eq. 2

$$V_n = V_m \sin\left(n\omega t\right) \tag{2}$$

Where, V_m is the rms voltage of any particular frequency (harmonic or power line).

The harmonics that are odd multiples of fundamental frequency are known as Odd harmonics and those that are even multiples of fundamental frequency are termed as Even harmonics. The frequencies that are in between the odd and even harmonics are called inter-harmonics.

Although, the ideal demand for any power utility is to have sinusoidal currents and voltages in AC system, this is not for all time promising, the currents and voltages with complex waveforms do occur in practice. Thus any complex waveform generated by such devices is a mixture of fundamental and the harmonics. Therefore, the voltage across a harmonically polluted system can be expressed numerically in eq. 3,

$$V = V_{fp}\sin\left(\omega t + \phi_1\right) + V_{2p}\sin\left(2\omega t + \phi_2\right) + V_{3p}\sin\left(3\omega t + \phi_3\right) + V_{np}\sin\left(n\omega t + \phi_n\right) \tag{3}$$

Where,

V_{fp} = Peak value of the fundamental frequency
V_{np}= Peak value of the nth harmonic component
ϕ = Angle of the respected frequency

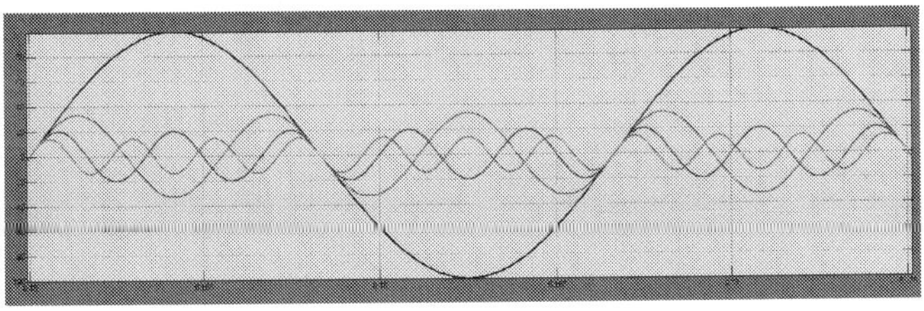

Figure 1. Fundamental and harmonics frequency waveforms

Similarly, the expression for current through a given circuit in a harmonically polluted system is given by the expression given in eq. 4

$$I = I_{fp}\sin(\omega t + \phi_1) + I_{2p}\sin(2\omega t + \phi_2) + I_{3p}\sin(3\omega t + \phi_3) \ldots\ldots + I_{np}\sin\left(n\omega t + \phi_n\right) \tag{4}$$

Harmonic components are also termed as positive, negative and zero sequence. In this case the harmonics that changes with the fundamental are called positive and those that have phasor direction opposite with the fundamental are called negative sequence components. The zero components do not take any affect from the fundamental and is considered neutral in its behavior. Phasor direction is pretty much important in case of motors. Positive sequence component tends to drive the motor in proper direction. Whereas the negative sequence component decreases the useful torque. The 7th, 13th, 19th etc. are positive sequence components. The negative sequence components are 5th, 11th, 17th and so on. The zero component harmonics are 3rd, 9th, 15th etc. As the amplitude of harmonics decreases with the increase in harmonic order therefore, in power systems the utilities are more concerned about the harmonics up to 11th order only.

3. Harmonics generation

In most of the cases the harmonics in voltage is a direct product of current harmonics. Therefore, the current harmonics is the actual cause of harmonics generation. Power line harmonics are generated when a load draws a non-linear current from a sinusoidal voltage. Nowadays all computers use Switch Mode Power Supplies (SMPS) that convert utility AC voltage to regulate low voltage DC for internal electronics. These power supplies have higher efficiency as compared to linear power supplies and have some other advantages too. But being based on switching principle, these non-linear power supplies draw current in high amplitude short pulses. These pulses are rich in harmonics and produce voltage drop across system impedance. Thus, it creates many small voltage sources in series with the main AC source as shown in Fig.2. Here in Fig.2 I_3 refers to the third harmonic component of the current drawn by the non-linear load, I_5 is the fifth harmonic component of the load current and so on. R shows the distributed resistance of the line and the voltage sources are shown to elaborate the factor explained above. Therefore, these short current pulses create significant distortion in the electrical current and voltage wave shape. This distortion in shape is referred as a harmonic distortion and its measurement is carried out in term of Total Harmonic Distortion (THD). This distortion travels back into the power source and can affect other equipment connected to the same source. Any SMPS equipment installed anywhere in the system have an inherent property to generate continuous distortion of the power source that puts an extra load on the utility system and the components installed in it. Harmonics are also produced by electric drives and DC-DC converters installed in industrial setups. Uninterrupted Power Supply (UPS) and Compact Fluorescent Lamp (CFL) are also a prominent source of harmonics in a system. Usually high odd harmonics results from a power electronics converter. In summary, the harmonics are produced in an electrical network by [2, 16, 26, 42]

- Rectifiers
- Use of iron core in power transformers
- Welding equipment
- Variable speed drives
- Periodic switching of voltage and currents
- AC generators by non-sinusoidal air gap, flux distribution or tooth ripple
- Switching devices like SMPS, UPS and CFL

It is worth mentioning here that voltage harmonics can emerge directly due to an AC generator, due to a non-sinusoidal air gap, flux distribution, or to tooth ripple, which is caused by the effect of the slots, which house the windings. In large supply systems, the greatest care is taken to ensure a sinusoidal output from the generator, but even in this case any non-linearity in the circuit will give rise to harmonics in the current waveform. Harmonics can also be generated due to the iron cores in the transformers. Such transformer cores have a non-linear B-H curve [37].

Figure 2. Voltage distortion due to non-linear current

4. Problems associated with harmonics

Harmonically polluted system has many threats for its stability. It not only hampers the power quality (PQ) but when a current is rich in harmonics, is drawn by some device, it overloads the system. For example third harmonic current has a property that unlike other harmonic component it adds up into the neutral wire of the system. This results in false tripping of circuit breaker. It also affects the insulation of the neutral cable. Overloading of the cables due to harmonically polluted current increases the losses associated with the wires. It should also be kept in mind that only the power from fundamental component is the useful power, rest all are losses. These additional losses make the power factor poor that results in more power losses. The overall summarized effects of harmonics in the power system include the following [9, 18, 39]

- Harmonic frequencies can cause resonant condition when combined with power factor correction capacitors
- Increased losses in system elements including transformers and generating plants
- Ageing of insulation
- Interruption in communication system
- False tripping of circuit breakers
- Large currents in neutral wires

The distribution transformers have a Δ-Y connection. In case of a highly third harmonic current the current that is trapped in the neutral conductor creates heat that increases the heat inside the transformer. This may lead to the reduced life and de-rating of transformer. The different types of harmonic have their own impact on power system. For instance let us consider the 3rd harmonic. Contrary to the balanced three phase system where the sum of all the three phases is zero in a neutral system, the third harmonic of all the three phases is identical. So it adds up in the neutral wire. The same is applicable on triple-n harmonics (odd multiples of 3 times the fundamental like 9th, 15th etc.). These harmonic currents are the main cause of false tripping and failure of earth fault protection relay. They also produce heat in the neutral wire thus a system needs a thicker neutral wire if it has third harmonic pollution in it. If a motor is supplied a voltage waveform with third harmonic content in it, it will only develop additional losses, as the useful power comes only from the fundamental component.

5. Harmonics monitoring standards

The identification of harmonics as a problem in AC power networks, has forced the utilities and regulatory authorities to devise the standards for harmonics monitoring and evaluation. The standards for harmonic control thus address both the consumers and the utility. Therefore, if the customer is not abiding by the regulations and is creating voltage distortion at the point of common coupling the utility can penalize him/her. Various renowned engineering institutes like IEEE, IEC and IET have devised laws to limit the injection of harmonic content in the grid. These standards are mostly helpful to achieve a user friendly healthy power quality system. IEEE standards are widely cited for their capability to address all the regions in the world. There are more than 1000 IEEE standards on electrical engineering fields. IEEE standards on power quality, however, are our main inspiration here. IEEE standard on harmonic control in electrical power system was published in 1992 and it covers all aspects related to harmonics [7]. It defines the maximum harmonics distortion up to 5 % on voltage levels ≤ 69kV. However, as the voltage levels are increased the allowable limits for harmonics in this standard are decreased to 1.5 % on all voltages ≥ 161 kV. It is also worth mentioning that individual voltage distortion starts from 3 % and ends at 1.0 % for voltage levels of ≤ 69kV and ≥ 161 kV respectively. Besides the standards that are designed keeping in view the global requirements, regional authorities devise their own standards according to their load profile and climatic conditions. Most of the standards are made according to the regional requirements of the country whereas few are based on the global needs and requirements. In Saudi Arabia there exists a regulatory body that defines the permissible limits and standard operational procedures for electricity transmission, distribution and generation. This body is known as electricity and co-generation regulatory authority [38]. Apart from devising standards they also follow some standards defined by UAE power distribution companies. One such standard defined by Saudi Electric Company (SEC) in 2007 and is known as "Saudi Grid Code". Harmonics limit set by the Saudi authorities is almost the same as IEEE standard but with a bit flexible limit of 3% THD for all networks operating within the range of 22kV-400kV [35, 38]. Table 1 compares the IEEE standard, the Abu Dhabi distribution company and the SEC standard for the harmonics limit in the electric network. It is interesting to mention that IEEE standard for controlling harmonics is silent for the conditions where a system is polluted with inter-harmonics (non-integer frequencies of fundamental frequency). For such conditions power utilities use IEC standard number 61000-2-2 .The IEC also defines the categories for different electronic devices in standard number 61000-3-2. These devices are then subjected to different allowable limits of THD. For example, class A has all three phase balanced equipment, non-portable tools, audio equipment, dimmers for only incandescent lamp. The limit for class A is varied according to the harmonic order. So for devices of class A the maximum allowable harmonic current is 1.08 A for 2nd, 2.3A for 3rd, 0.43A for 4th, 1.14A for 5th harmonics. The beauty of this IEC standard is that it also caters for power factor. For example all devices of class C (lighting equipment other than the incandescent lamp dimmer) have 3rd harmonic current limit as a function of circuit power factor.

	SEC Standard	Abu Dhabi Distribution Company	IEEE Limits
Harmonics	THD limit is 5% for 400 V system, and 4% and 3% for 6.6-20kV and 22kV-400kV respectively	THD limit is 5% for 400 V system, and 4% and 3% for 6.6-20kV and 22kV-400kV respectively	5% for all voltage levels below 69kV and 3% for all voltages above 161 kV

Table 1. Comparison of Harmonic Standards [7, 35, 38]

The modern systems based on artificial intelligent techniques like Fuzzy logic, ANFIS and CI based computations are reducing the difficulty of data mining that helps in redesigning the standards for power quality harmonics [24, 25]. In developed countries like Australia, Canada, USA the power distribution companies are already partially shifted to smart grid and they are using sophisticated sensors and measuring instruments.

In terms of smart grid environment these sensors will help in mitigating the problems by predicting them in advance. Smart grid, by taking intelligent measurements and by the aid of sophisticated algorithms will be able to predict the PQ problems like harmonics, fault current in advance. It is pertinent to mention that the power quality monitoring using the on-going 3G technologies has been implemented by Chinese researchers. They used module of GPRS that is capable of analyzing the real time data and its algorithm makes it intelligent enough to get the desired PQ information [22].

6. Harmonics measurement

The real challenge in a harmonically polluted environment is to understand and designate the best point for measuring the harmonics. Nowadays the revolution in electronics has messed up the AC system so much that almost every user in a utility is a contributor to the harmonics current. Furthermore, the load profile in any domestic area varies from hour to hour within a day. So in order to cope with the energy demand and to improve the power factor, utilities need to switch on and off the power factor correction capacitors. This periodic and non-uniform switching also creates harmonics in the system. The load information in an area although, provide some basic information about the order of harmonic present in a system. Such information is very useful as it gives a bird eye view of harmonic content. But for the exact identification of the harmonics it is necessary to synthesize the distorted waveform using the power quality analyzer or using some digital oscilloscope for Fast Fourier Transform (FFT). For example Fig.3 shows a general synthesis of the current drawn by a controlled rectifier. Once identified, the level and type of harmonics (3rd, 5th etc.) the steps to mitigation can be devised. It should be kept in mind that proper measurement is the key for the proper designing of harmonic filters. But the harmonics level may differ at different points of measurement in a system. Therefore, utilities need to be very precise in identifying the correct point for harmonic measurement in a system. Among the standards, it is IEEE standard 519-1992 that outlines the operational procedures for carrying out the harmonic measurements. This standard however does not state any restriction regarding the integration duration of the measurement equipment with the system. It however, restricts the utility to maintain a log for monthly records of maximum demand [5]. Various devices are used in support with each other to carry out the harmonic measurements in a system. These include the following

- Power Quality Analyser
- Instrument transformers based transducers (CT and PT)

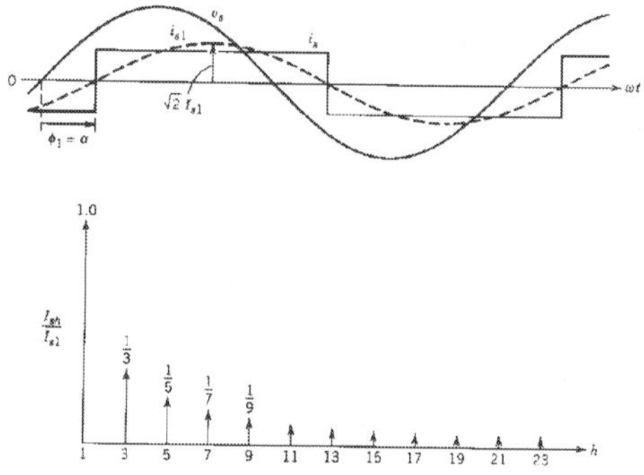

Figure 3. Typical line current of a controlled converter [26]

Various renowned companies are designing and producing excellent PQ analyzers. These include FLUKE, AEMC, HIOKI, DRANETZ and ELSPEC. These companies design single phase and three phase PQ analyzers that are capable of measuring all the dominant harmonic frequencies. The equipment that is used for harmonic measurement is also bound to some limitations for proper harmonic measurement. This limitation is technical in nature as for accurate measurement of all harmonic currents below the 65th harmonic, the sampling frequency should be at least twice the desired input bandwidth or 8k samples per second in this case, to cover 50Hz and 60Hz systems [5]. Mostly, the PQ analyzers are supplied along with the CT based probes but depending on the voltage and current ratings a designer can choose the CT and PT with wide operating frequency range and low distortion. The distance of equipment with the transducer is also very important in measuring harmonics. If the distance is long then noise can affect the measurement therefore properly shielded cables like coaxial cable or fiber optic cables are highly recommended by the experts [5]. In short, the measurement of harmonics should be made on Point of Common Coupling (PCC) or at the point where non-linear load is attached. This includes industrial sites in special as they are the core contributors in injecting harmonic currents in the system.

7. Harmonics purging techniques

Techniques have been designed and tested to tackle this power quality issue since the problem is identified by the researchers. There are several techniques in the literature that addresses the mitigation of harmonics. All these techniques can be classified under the umbrella of following

i. Passive harmonic filter
ii. Active harmonic filter
iii. Hybrid harmonic filter
iv. Switching techniques

7.1. Passive harmonic filters

Passive filter techniques are among the oldest and perhaps the most widely used techniques for filtering the power line harmonics. Besides the harmonics reduction passive filters can be used for the optimization of apparent power in a power network. They are made of passive elements like resistors, capacitors and inductors. Use of such filters needs large capacitors and inductors thus making the overall filter heavier in weight and expensive in cost. These filters are fixed and once installed they become part of the network and they need to be redesigned to get different filtering frequencies. They are considered best for three phase four wire network [18]. They are mostly the low pass filter that is tuned to desired frequencies. Giacoletto and Park presented an analysis on reducing the line current harmonics due to personal computer power supplies [10]. Their work suggested that the use of such filters is good for harmonics reduction but this will increase the reactive component of line current. Various kind of passive filter techniques are given below [18, 19].

i. Series passive filters
ii. Shunt passive filters
iii. Low pass filters or line LC trap filters
iv. Phase shifting transformers

7.1.1. Series passive filters

Series passive filters are kinds of passive filters that have a parallel LC filter in series with the supply and the load. Series passive filter shown in Fig.4 are considered good for single phase applications and specially to mitigate the third harmonics. However, they can be tuned to other frequencies also. They do not produce resonance and offer high impedance to the frequencies they are tuned to. These filters must be designed such that they can carry full load current. These filters are maintenance free and can be designed to significantly high power values up to MVARs [4]. Comparing to the solutions that employ rotating parts like synchronous condensers they need lesser maintenance.

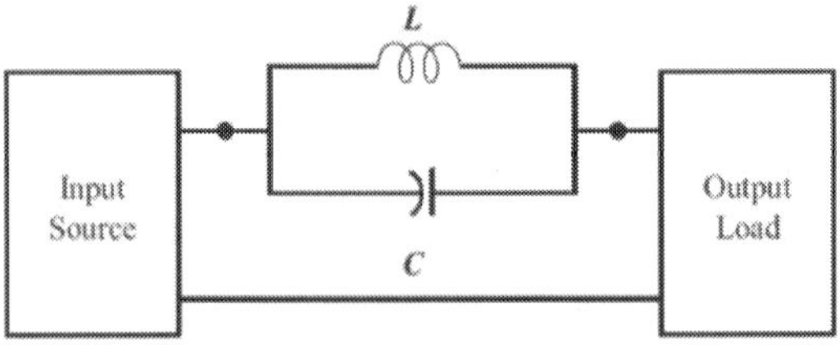

Figure 4. Passive Series Filter [18]

7.1.2. Shunt passive filters

These type of filters are also based on passive elements and offer good results for filtering out odd harmonics especially the 3rd, 5th and 7th. Some researchers have named them as single tuned filters, second order damped filters and C type damped filters [3]. As all these filters come in shunt with the line they fall under the cover of shunt passive filters, as shown in Fig.5. Increasing the order of harmonics makes the filter more efficient in working but it reduces the ease in designing. They provide low impedance to the frequencies they are tuned for. Since they are connected in shunt therefore they are designed to carry only harmonic current [18]. Their nature of being in shunt makes them a load itself to the supply side and can carry 30-50% load current if they are feeding a set of electric drives [13]. Economic aspects reveal that shunt filters are always economical than the series filters due to the fact that they need to be designed only on the harmonic currents. Therefore they need comparatively smaller size of L and C, thereby reducing the cost. Furthermore, they are not designed with respect to the rated voltage, thus makes the components lesser costly than the series filters [33]. However, these types of filters can create resonant conditions in the circuit.

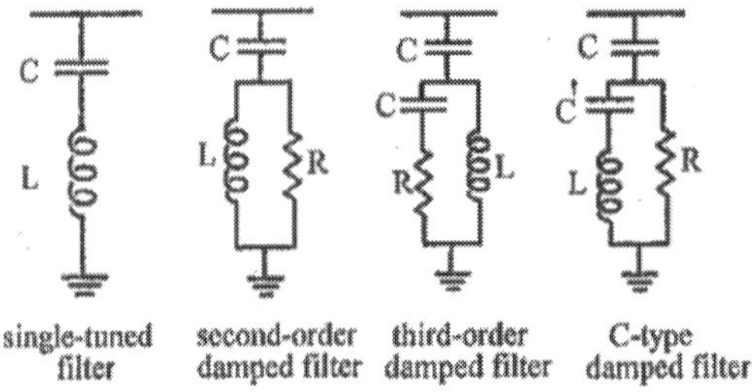

Figure 5. Different order type shunt filters [3]

7.1.3. Low pass filter

Low pass filters are widely used for mitigation of all type of harmonic frequencies above the threshold frequency. They can be used only on nonlinear loads. They do not pose any threats to the system by creating resonant conditions. They improve power factor but they must be designed such that they are capable of carrying full load current. Some researchers have referred them as line LC trap filters [19]. These filters block the unwanted harmonics and allow a certain range of frequencies to pass. However, very fine designing is required as far as the cut off frequency is concerned.

7.1.4. Phase shifting transformers

The nasty harmonics in power system are mostly odd harmonics. One way to block them is to use phase shifting transformers. It takes harmonics of same kind from several sources in a network and shifts them alternately to 180° degrees and then combine them thus resulting in cancelation. We have classified them under passive filters as transformer resembles an inductive network. The use of phase shifting transformers has produced considerable success in suppressing harmonics in multilevel hybrid converters [34]. S. H. H. Sadeghi et.al. designed an algorithm that based on the harmonic profile incorporates the phase shift of transformers in large industrial setups like steel industry [36].

7.2. Active harmonic filters

In an Active Power Filter (APF) we use power electronics to introduce current components to remove harmonic distortions produced by the non-linear load. Figure 6 shows the basic concept of an active filter [27]. They detect the harmonic components in the line and then produce and inject an inverting signal of the detected wave in the system [27]. The two driving forces in research of APF are the control algorithm for current and load current analysis method [23]. Active harmonic filters are mostly used for low-voltage networks due to the limitation posed by the required rating on power converter [21].

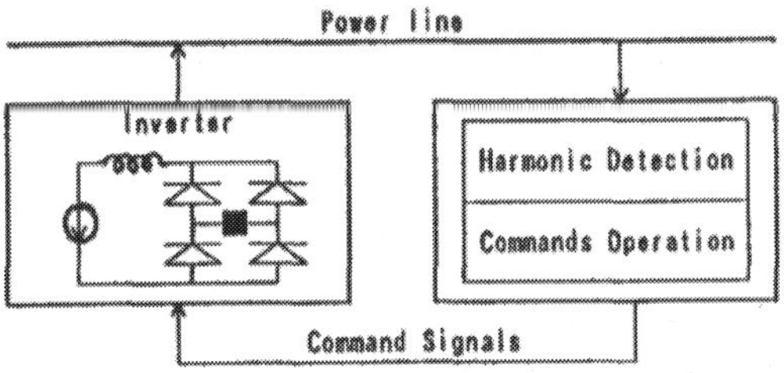

Figure 6. Conceptual demonstration of Active filter [27]

They are used even in aircraft power system for harmonic elimination [6]. Same like passive filters they are classified with respect to the connection method and are given below [40].

i. Series active filters
ii. Shunt active filters

Since, it uses power electronic based components therefore in literature a lot of work has been done on the control of active filters.

7.2.1. Series active filter

The series filter is connected in series with the ac distribution network as show in Fig.7 [33]. It serves to offset harmonic distortions caused by the load as well as that present in the AC system. These types of active filters are connected in series with load using a matching transformer. They inject voltage as a component and can be regarded as a controlled voltage source [33]. The drawback is that they only cater for voltage harmonics and in case of short circuit at load the matching transformer has to bear it [31].

7.2.2. Shunt active filter

The parallel filter is connected in parallel with the AC distribution network. Parallel filters are also known as shunt filters and offset the harmonic distortions caused by the non-linear load. They work on the same principal of active filters but they are connected in parallel as stated that is they act as a current source in parallel with load [21]. They use high computational capabilities to detect the harmonics in line.

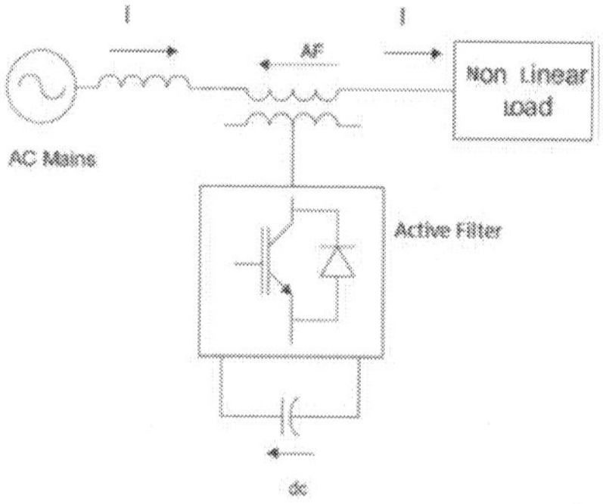

Figure 7. Series active filters [33]

Mostly microprocessor or micro-controller based sensors are used to estimate harmonic contents and to decide the control logic. Power semiconductor devices are used especially the IGBT. Some researchers claim that before the advent of IGBTs active filters were seldom use due to overshoot in budget [11]. However, despite of their usefulness shunt active filters have many drawbacks. Practically they need a large rated PWM inverter with quick response against system parameters changes. If the system has passive filters attached somewhere, as in case of hybrid filters then the injected currents may circulate in them [28].

7.3. Hybrid harmonic filters

These types of filters combine the passive and active filters. They contain the advantages of active filters and lack the disadvantages of passive and active filters. They use low cost high power passive filters to reduce the cost of power converters in active filters that is why they are now very much popular in industry. Hybrid filters are immune to the system impedance, thus harmonic compensation is done in an efficient manner and they do not produce the resonance with system impedance [29]. The control techniques used for these types of filters are based on instantaneous control, on p-q theory and i_d-i_q. K.N.M.Hasan et.al. presented a comparative study among the p-q and i_d-i_q techniques and concluded that in case of voltage distortions the i_d-i_q method provides slightly better results [12]. They are usually combined in the following ways [21]

i. Passive series active series hybrid filters
ii. Passive series active shunt hybrid filters
iii. Passive shunt active series hybrid filters
iv. Passive shunt active shunt hybrid filters

7.3.1. Passive series active series hybrid filters

These type of hybrid filters have both kind of filters connected in series with the load as shown in Fig.8 and are considered good for diode rectifiers feeding a capacitive load [32]

7.3.2. Passive series active shunt hybrid filters

This breed of hybrid filter has passive part in series with load and active filter in parallel. AdilM. Al-Zamil et al. proposed such type of filters in their paper and used the high power capability. of passive filter by placing them in series with the load. They used an active filter with space vector pulse with modulation (SVPWM) and implemented it on micro-controller. They used only line current sensors to compute all the parameters required for reference current generation. Their proposed system worked satisfactorily up to the 33rd harmonic and the results shown are based on a system with line reactance of 0.13 pu. In their system the bandwidth required for active filter is relatively less due to the passive filter that takes care of the rising and falling edges of load current. They proposed that while designing hybrid system the line filter L and capacitance C of active filter needs a compromise in selection depending on the acceptable level of switching frequency ripple current and minimum acceptable ripple voltage [1].

7.3.3. Passive shunt active shunt hybrid filters

These types of filters have both the passive and active filters connected in shunt with the load as shown in Fig.9 [21]. In a comparative study J.Turunen et al. claimed that they require smallest transformation ratio of coupling transformer as a result they need a fairly high power rating for a small load and in case of high power loads the problem of dc link control results in poor current filtering [43].

7.3.4. Passive shunt active series hybrid filters

As its name implies it is a kind of hybrid filter that has an active filter in series and a passive filter in shunt as shown in Fig.10. J. Turunen et al. in a comparative study stated that this breed of hybrid filter utilizes very small transformation ratio therefore for same rating of load their power rating required is large compared to the load [43].

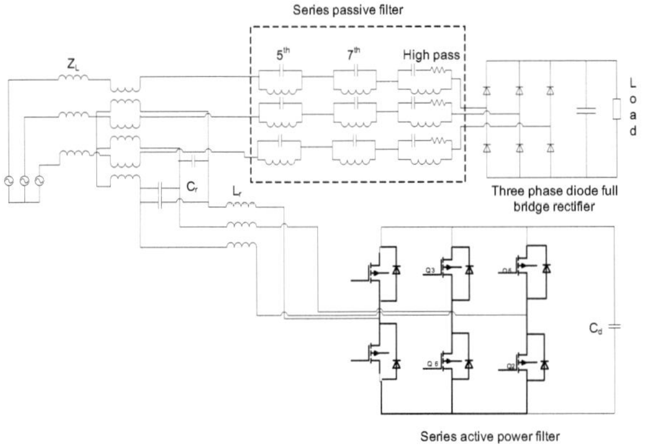

Figure 8. Passive series active series hybrid filters [32]

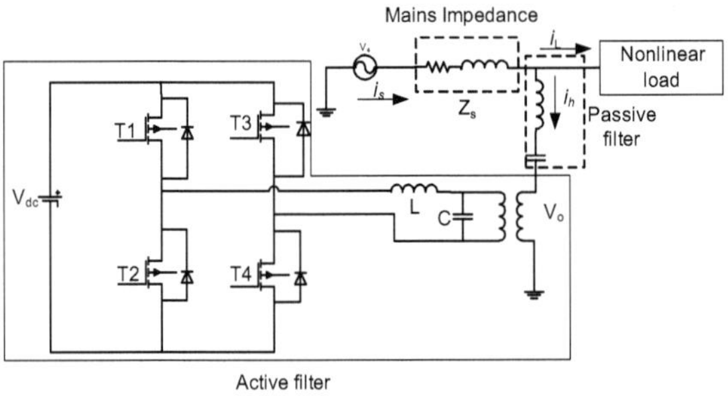

Figure 9. Passive shunt active shunt hybrid filters [21]

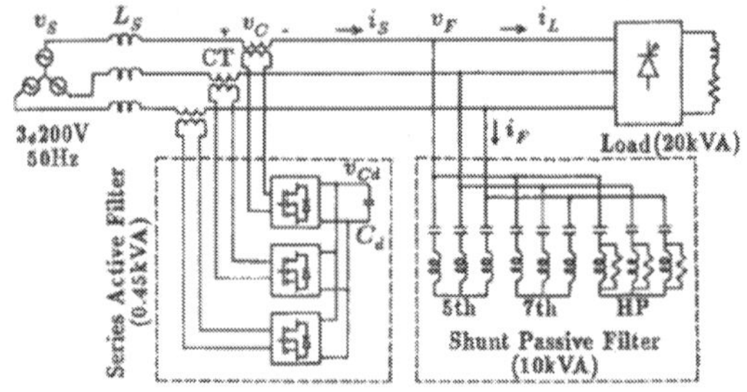

Figure 10. Active series passive shunt hybrid filters [29]

7.4. Switching techniques

Besides using the method of installing filters, power electronics is so versatile that up to some extent harmonics can be eliminated using switching techniques. These techniques may vary from the increasing the pulse number to advance algorithm based Pulse Width

Modulation (PWM). The most widely used sine triangle PWM was proposed in 1964. Later in 1982 Space Vector PWM (SVPWM) was proposed [20]. PWM is a magical technique of switching that gives unique results by varying the associated parameters like modulation index, switching frequency and the modulation ratio. The frequency modulation ratio '*m*' if taken as odd automatically removes even harmonics [17, 26]. Here the increase in switching frequency reduces the current harmonics but this makes the switching losses too much. Furthermore, we cannot keep on increasing switching frequency because this imposes the EMC problems [15]. D.G.Holmes et al. presented an analysis for carrier based PWM and claimed that it is possible to use some analytical solutions to pin point the harmonic cancelation using different modulation techniques. Sideband harmonics can be eliminated if the designer uses natural or asymmetric regular sampled PWM [14]. The output can be improved by playing with the modulation index. One specialized type of PWM is called Selective Harmonic Elimination (SHE) PWM or the programmed harmonic elimination scheme. This technique is based on Fourier analysis of phase to ground voltage. It is basically a combination of square wave switching and the PWM. Here proper switching angles selection makes the target harmonic component zero [26, 30]. In SHE technique a minimum of 0.5 modulation index is possible [41]. But even the best SHE left the system with some unfiltered harmonics. J. Pontt et al. presented a technique of treating the unfiltered harmonics due to the SHE PWM. They stated that if we use SHE PWM for elimination of 11th and 13th harmonics for 12 pulse configuration then the harmonics of order 23th, 25th, 35th and 37th are one that play vital role in defining the voltage distortions. They proposed the use of three level active front end converters. They suggested a modulation index of 0.8-0.98 to mitigate the harmonics of order 23rd, 25th and 35th, 37th [30]. With some modifications researchers have shown that SHE PWM can be used at very low switching frequency of 350 Hz. Javier Napoles et al. presented this technique and give it a new name of Selective Harmonic Mitigation (SHM) PWM. They used seven switching states and results makes the selective harmonics equal to zero [8]. This is excellent since in SHE PWM the selective harmonic need not to be zero. It is sufficient in conventional PWM to bring it under the allowable limit. Siriroj Sirisukprasert et al. presented an optimal harmonic reduction technique by varying the nature of output stepped waveforms and varied the modulation indexes. They tested their proposed technique on multilevel inverters that are better than the two level conventional inverters. They excluded the very narrow and very wide pulses from the switching waveform. Unlike SHE PWM as discussed above they ensured the minimum turn on and turn off by switching their power switches only once a cycle. Contrary to traditional SHE PWM, in this case the modulation index can vary till 0.1. The output is a stepped waveform for different stages they classify the production of modulation index as high, low and medium and the real point of interest is that for all these three classes of modulation indexes the switching is once per cycle per switch [41]. Some researchers used trapezoidal PWM method for harmonic control. This kind of PWM is based on unipolar PWM switching. Here a trapezoidal waveform is compared with a triangular waveform and the resulting PWM is supplied to the power switches. Like other harmonic elimination techniques in PWM based techniques researchers have proposed the use of AI based techniques including FL and ANN.

8. Conclusion

This chapter summarizes one of the major power quality problems that is the reason of many power system disturbances in an electrical network. The possible sources of harmonics are discussed along with their effects on distribution system components including the transformers, switch gears and the protection system. The regulatory standards for the limitation of harmonics and their measurement techniques are also presented here. The purging techniques of harmonics are also presented and various kind of harmonic filters are briefly presented. To strengthen the knowledge base, this chapter has also discussed the control of harmonics using PWM techniques. By this chapter we have attempted to gather the technical information in this field. A thorough understanding of harmonics will provide the utility engineers a framework that is often required in the solution of research work related to harmonics.

Author details

Hadeed Ahmed Sher* and Khaled E Addoweesh
Department of Electrical Engineering, King Saud University, Riyadh, Saudi Arabia

Yasin Khan
Department of Electrical Engineering, King Saud University, Riyadh, Saudi Arabia
Saudi Aramco Chair in Electrical Power, Department of Electrical Engineering, King Saud University, Riyadh

9. References

[1] A.M. Al-Zamil and D.A. Torrey. "A passive series, active shunt filters for high power applications". Power Electronics, IEEE Transactions on, 16(1):101–109, 2001.
[2] S.J.Chapman. "Electric machinery fundamentals". McGraw-Hill Science/ Engineering/Math, 2005.
[3] C.J. Chou, C.W. Liu, J.Y. Lee, and K.D. Lee. "Optimal planning of large passive-harmonic filters set at high voltage level". Power System IEEE Transactions on, 15(1):433–441, 2000.
[4] JC Das. "Passive filters-potentialities and limitations". In Pulp and Paper Industry Technical Conference, 2003. Conference Record of the 2003 Annual, pages 187–197. IEEE, 2003.
[5] F. De la Rosa and Engnetbase. "Harmonics and power systems". Taylor&Francis, 2006.
[6] A. Eid, M. Abdel-Salam, H. El-Kishky, and T. El-Mohandes. "Active power filters for harmonic cancellation in conventional and advanced aircraft electric power systems." Electric Power Systems Research, 79(1):80–88, 2009.
[7] I. F II. "IEEE recommended practices and requirements for harmonic control in electrical power systems". 1993.
[8] L.G. Franquelo, J. Napoles, R.C.P. Guisado, J.I. Leon, and M.A. Aguirre. "A flexible selective harmonic mitigation technique to meet grid codes in three-level PWM converters". Industrial Electronics, IEEE Transactions on, 54(6):3022–3029, 2007.

* Corresponding Author

[9] E.F. Fuchs and M.A.S. Masoum. "Power quality in power systems and electrical machines" .Academic Press, 2008.
[10] LJ Giacoletto and GL Park. "Harmonic filtering in power applications". In Industrial and Commercial Power Systems Technical Conference, 1989, Conference Record., pages 123–128. IEEE, 1989.
[11] C.A. Gougler and JR Johnson. "Parallel active harmonic filters: economical viable technology". In Power Engineering Society 1999 Winter Meeting, IEEE, volume 2, pages 1142–1146. IEEE.
[12] K.N.M. Hasan and M.F. Romlie. "Comparative study on combined series active and shunt passive power filter using two different control methods". In Intelligent and Advanced Systems, 2007. ICIAS 2007. International Conference on, pages 928–933. IEEE, 2007.
[13] F.L. Hoadley. "Curb the disturbance". Industry Applications Magazine, IEEE, 14(5):25–33,2008.
[14] D.G. Holmes and B.P. McGrath. "Opportunities for harmonic cancellation with carrier-based PWM for a two-level and multilevel cascaded inverters." Industry Applications, IEEE Transactions on, 37(2):574–582, 2001.
[15] J. Holtz. "Pulse width modulation-A survey." Industrial Electronics, IEEE Transactions on, 39(5):410–420, 1992.
[16] H.Rashid. "Power Electronics Circutis Devices and Applications". Prentice Hall Int. Ed.,1993.
[17] I.B. Huang and W.S. Lin. "Harmonic reduction in inverters by use of sinusoidal pulsewidth modulation." Industrial Electronics and Control Instrumentation, IEEE Transactions on, (3):201–207, 1980.
[18] J. David Irwin. "The industrial electronics handbook". CRC, 1997.
[19] D. Kampen, N. Parspour, U. Probst, and U. Thiel. "Comparative evaluation of passive harmonic mitigating techniques for six pulse rectifiers" In Optimization of Electrical and Electronic Equipment, 2008. OPTIM 2008. 11th International Conference on, pages 219–225. IEEE, 2008.
[20] M.P. Ka'zmierkowski and R. Krishnan. "Control in power electronics: selected problems." Academic Pr, 2002.
[21] B.R. Lin, B.R. Yang, and H.R. Tsai. "Analysis and operation of hybrid active filter for harmonic elimination". Electric Power Systems Research, 62(3):191–200, 2002.
[22] D. LU, H. ZHANG, and C. WANG. "Research on the reliable data transfer based on udp" [j]. Computer Engineering, 22, 2003.
[23] L. Marconi, F. Ronchi, and A. Tilli. "Robust nonlinear control of shunt active filters for harmonic current compensation". Automatica, 43(2):252–263, 2007.
[24] W.G. Morsi and ME El-Hawary. "A new fuzzy-based representative quality power factor for unbalanced three-phase systems with non-sinusoidal situations" Power Delivery, IEEE Transactions on, 23(4):2426–2438, 2008.
[25] S. Nath and P. Sinha. "Measurement of power quality under non-sinusoidal condition using wavelet and fuzzy logic". In Power Systems, 2009. ICPS'09. International Conference on, pages 1–6. IEEE,2009.
[26] N. Mohan,T. Undeland and W. P. Robbins. "Power Electronics Converters, Applications and Design" Wiley India, 2006.

[27] N. Pecharanin, M. Sone, and H. Mitsui. "An application of neural network for harmonic detection in active filter". In Neural Networks, 1994. IEEE World Congress on Computational Intelligence., 1994 IEEE International Conference on, volume 6, pages 3756–3760. IEEE, 1994.

[28] F.Z. Peng, H. Akagi, and A. Nabae. "A new approach to harmonic compensation in power systems-a combined system of shunt passive and series active filters". Industry Applications, IEEE Transactions on, 26(6):983–990, 1990.

[29] F.Z. Peng, H. Akagi, and A. Nabae. "Compensation characteristics of the combined system of shunt passive and series active filters". Industry Applications, IEEE Transactions on, 29(1):144–152, 1993.

[30] J. Pontt, J. Rodriguez, R. Huerta, and J. Pavez. "A mitigation method for non-eliminated harmonics of SHE PWM three-level multipulse three-phase active front end converter." In Industrial Electronics, 2003. ISIE'03. 2003 IEEE International Symposium on, volume 1, pages 258–263. IEEE, 2003.

[31] NA Rahim, S. Mekhilef, and I. Zahrul. "A single-phase active power filter for harmonic compensation". In Industrial Technology, 2005. ICIT 2005. IEEE International Conference on, pages 1075–1079. IEEE, 2005.

[32] S. Rahmani, K. Al-Haddad, and F. Fnaiech. "A hybrid structure of series active and passive filters to achieving power quality criteria". In Systems, Man and Cybernetics, 2002 IEEE International Conference on, volume 3, pages 6–pp. IEEE, 2002.

[33] M.H. Rashid. "Power electronics handbook". Academic Pr, 2001.

[34] C. Rech and JR Pinheiro. "Line current harmonics reduction in hybrid multilevel converters using phase-shifting transformers". In Power Electronics Specialists Conference, 2004. PESC04. 2004 IEEE 35th Annual, volume 4, pages 2565–2571. IEEE, 2004.

[35] Regulation, supervision bureau for the water, and electricity sector of the Emirate of Abu Dhabi. "Limits for harmonics in the electricity supply system". 2005.

[36] SHH Sadeghi, SM Kouhsari, and A. Der Minassians. "The effects of transformers phase-shifts on harmonic penetration calculation in a steel mill plant". In Harmonics and Quality of Power, 2000. Proceedings. Ninth International Conference on, volume 3, pages 868–873. IEEE, 2000.

[37] C. Sankaran. "Power quality". CRC, 2002.

[38] SEC. "The Saudi Arabian grid code", 2007.

[39] J. Shepherd, A.H. Morton, and L.F. Spence. "Higher electrical engineering." Pitman Pub.,1975.

[40] B. Singh, K. Al-Haddad, and A. Chandra. "A review of active filters for power quality improvement". Industrial Electronics, IEEE Transactions on, 46(5):960–971, 1999.

[41] S. Sirisukprasert, J.S. Lai, and T.H. Liu. "Optimum harmonic reduction with a wide range of modulation indexes for multilevel converters". Industrial Electronics, IEEE Transactions on, 49(4):875–881, 2002.

[42] W. Theodore et al. "Electrical Machines, Drives And Power Systems" 6/E. Pearson Education India, 2007.

[43] J. Turunen, M. Salo, and H. Tuusa. "Comparison of three series hybrid active power filter topologies" In Harmonics and Quality of Power, 2004. 11th International Conference on, pages 324–329. IEEE, 2004.

Chapter 2

Power Quality and Grid Code Issues in Wind Energy Conversion System

Sharad W. Mohod and Mohan V. Aware

Additional information is available at the end of the chapter

http://dx.doi.org/10.5772/54704

1. Introduction

The aim of the electric power system is to produce and deliver to the consumer's electric energy of defined parameters, where the main quantities describing the electric energy are the voltage and frequency. During normal operation of system the frequency varies as a result of the variation of the real power generated and consumed. At the same time, because of voltage drops in the transmission lines and transformers it is impossible to keep the voltage at the nominal level in all the nodes of the power system. It is also impossible to keep an ideal sinusoidal shape of the voltage or current waveform due to the nonlinearities in many devices use for electric energy generation, transmission and at end users. That is why the electric power system require to keep the quantities near the nominal value[1]-[5].

Recently, the deregulated electricity market has also opened the door for customers own distributed generation due to economical and technical benefit. The liberalization of the grid leads to new management structures, in which the trading of energy is important. The need to integrate the renewable energy like wind energy into power system is to minimize the environmental impact on conventional plant of generation. The conventional plant uses fossil fuels such as coal & petroleum products to run the steam turbines and generate the thermal power. The fossil fuel consumption has an adverse effect on the environment and it is necessary to minimize the polluting and exhausting fuel. The penetration of renewable energy especially wind has been increasing fast during the past few years and it is expected to rise more in near future. Many countries around the world are likely to experience similar penetration level. During the last decade of the twentieth century, worldwide wind energy capacity is doubled approximately every three years.

Today's trends are to connect all size of generating units like wind farm,solar farm,biogas generation and conventional source like coal,hydro,nuclear power plant in to the grid system shown in Fig.1.0

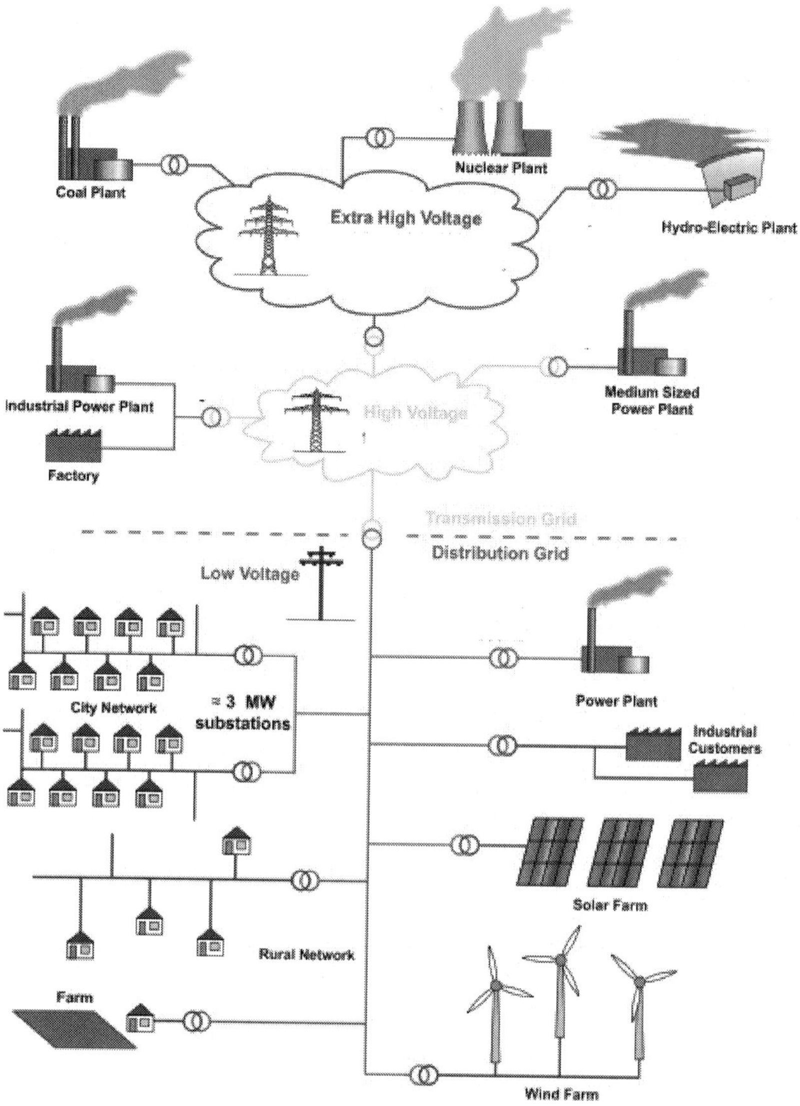

Figure 1. Grid integration of interconnected system

The critical power quality issues related to integration of wind farms have been identified by team of Riso National Laboratory and Danish Utility Research Institute, Denmark and Electronic Research and Development Centre, India in Nov.1998 .The power quality in relation to a wind turbine describe the electrical performance of wind energy generating system. It reflects the generation of grid interference and the influence of a wind turbine on power and voltage quality of grid. The issue of power quality is of great importance to the wind turbines. There has been an extensive growth and quick development in the exploitation of wind energy in recent years [6]-[7]. The individual units can be of large capacity up to 5 MW, feeding into distribution network, particularly with customers connected in close proximity. However with rapidly varying voltage fluctuations due to the nature of wind, it is difficult to improve the power quality with simple compensator. Advance reactive power compensators with fast control and power electronic have emerged to supersede the conventional reactive compensator [8]-[9].

It has been suggested that today's industrial development are related with generalized use of computers, adjustable speed drives and other microelectronic loads. It also becomes an increasing concern with power quality to the end customer. The presence of harmonic and reactive power in the grid is harmful, because it will cause additional power losses and malfunction of grid component. The massive penetration of electronically controlled devices and equipments in low voltage distribution network is responsible for further worsening of power-quality problem [10]-[13].

The problems are related to the load equipment and devices used in electric energy generation. Now a days the transmission and distribution system become more sensitive to power quality variation than those used in the past. Many new devices contain microprocessor based controls and electronics power elements that are sensitive to many types of disturbances. The wind turbine generating systems are the highly variable sources of energy and wind turbine are belonging to the source of such problem.

The wind power in the electric grid system affects the voltage quality. To assess this effect, the knowledge of about the electrical characteristic of wind turbine is needed. The electrical characteristics of wind turbine are manufacturer's specification and not site specification. This means that by having the actual parameter values for a specific wind turbine the expected impact of the wind turbine on voltage quality is important. The need for consistent and replicable documentation of the power quality characteristics of wind turbines, the International Electro-technical Commission (IEC) started work to facilitate for power quality in 1996. As a result, IEC 61400-21 was developed and today most wind turbines manufacturers provide power quality characteristic data accordingly. Wind turbines and their power quality will be certified on the basis of measurements according to national or international guidelines. These certifications are an important basis for utilities to evaluate the grid connection of wind turbines and wind farms.

The power quality is defined as set of parameters defining the properties of the power supply as delivered to user in normal operating condition in terms of continuity of supply and characteristics of voltage, frequency.

Today the measurement and assessment of the power quality characteristics of the grid-connected wind turbines is defined by IEC Standard 61400-21 (wind turbine system) prepared by IEC- Technical Committee 88.

The need of power quality in wind integration system and its issues are highlighted in further section.

2. Need of power quality studies

The power quality studies are of importance to wind turbine as a individual units can be large up to 5 MW, feeding into distribution circuit with high source impedance and with customer connected in close proximity.

With the advancement in fast switching power devices there is a trend for power supply size reduction. The current harmonics due to switching converters makes supply current distorted. The increase of electronic controllers in drives, furnaces, household equipments and SMPS are increasing the harmonic content and reactive power in electric supply. The distribution transformers apart from reactive loads draw reactive current from the supply to meet the magnetizing current. The ever-increasing demand for power is not fulfilled by increase in generation and particularly in distribution for various reasons such as environmental issues, increasing cost of natural fuel, opposition to nuclear power plants, etc. This puts excessive burden on the electric supply resulting in poor power quality. The term power quality here refers to the variation in supply voltage, current and frequency. The excessive load demand tries to retard the turbines at generation plant. This results in reduction in voltage and more severely reduction in the supply frequency. The authorities are working for power quality improvement by using reactive compensators and active filters on supply side and penalizing consumers for polluting the power grid.

The increasing problems and advances in power electronic technology, has forced to change the traditional power system concepts. Use of fast reactive power compensators can improve the power system stability and hence, the maximum power transfers through the electric system.

The reactive power in its simpler form, for a single phase sinusoidal voltages and current can be defines as the product of a phase current (reactive component) and the supply voltage. There is a simple right angle triangle relation between active power, reactive power and apparent power. But, this definition of the reactive power is not sufficient for non-linear loads where fundamental current and fundamental voltage may not have any phase difference. However, for such loads, power factor is still less than unity. The power factor definition is modified to accommodate for non-linear loads.

The overall power factor has two parts, the displacement power factor and distortion power factor. The displacement power factor defined as cosine of phase shift between fundamental supply current and voltage.

Distortion power factor "DF" or harmonic factor is defined as the ratio of the RMS harmonic content to the RMS value of fundamental component expressed as percentage of the fundamental.

$$DF = \sqrt{\frac{\text{sum of squares of amplitudes of all harmonics}}{\text{square of amplitude of fundamental}}} * 100\% \quad (1)$$

$$DF(for\ \text{current}) = \frac{\sqrt{\sum_{h=2}^{\infty} I_h^2}}{I_1} \quad (2)$$

2.1. Issue of voltage variation

If a large proportion of the grid load is supplied by wind turbines, the output variations due to wind speed changes can cause voltage variation, flicker effects in normal operation. The voltage variation can occur in specific situation, as a result of load changes, and power produce from turbine. These can expected in particular in the case of generator connected to the grid at fixed speed. The large turbine can achieve significantly better output smoothing using variable speed operation, particularly in the short time range. The speed regulation range is also contributory factor to the degree of smoothing with the large speed variation capable of suppressing output variations.

2.2. Issue of voltage dips

It is a sudden reduction in the voltage to a value between 1% & 90 % of the nominal value after a short period of time, conventionally 1ms to 1 min. This problem is considered in the power quality and wind turbine generating system operation and computed according to the rule given in IEC 61400-3-7 standard, "Assessment of emission limit for fluctuating load". The start up of wind turbine causes a sudden reduction of voltage. The relative % voltage change due to switching operation of wind turbine is calculated as

$$d = 100K_u(\Psi_k)\frac{S_n}{S_k^*} \quad (3)$$

Where d - Relative voltage change,

$k_u(\Psi_k)$ - Voltage change factor,

S_n - Rated apparent power of wind turbine and S_k^* short circuit apparent power of grid. The voltage dips of 3% in most of the cases are acceptable. When evaluating flicker and power variation within 95% of maximum variation band corresponding to a standard deviation are evaluated.

2.3. Switching operation of wind turbine on the grid

Switching operations of wind turbine generating system can cause voltage fluctuations and thus voltage sag, voltage swell that may cause significant voltage variation. The acceptances of switching operation depend not only on grid voltage but also on how often this may occur. The maximum number of above specified switching operation within 10-minute period and 2-hr period are defined in IEC 61400-3-7 Standard.

Voltage sag is a phenomenon in which grid voltage amplitude goes below and then returns to the normal level after a very short time period. Generally, the characteristic quantity of voltage sag is described by the amplitude and the duration of the sags. The IEEE power quality standards define the voltage sag when the amplitude of voltage is 0.1–0.9 p.u. value and its duration is between 10 ms and 1 min. A voltage sag is normally caused by short-circuit faults in the power network or by the starting up of Induction Generator/Motors.

The bad weather conditions, such as thunderstorm, single-phase earthed faults are the causes of voltage sags. In addition, large electric loads such as large electrical motors or arc furnaces can also cause voltage sags during the startup phase with serious current distortion.

The adverse consequences are the reduction in the energy transfer of electric motors. The disconnection of sensitive equipments and thus the industrial process may bring to a standstill.

2.4. Harmonics

The harmonics distortion caused by non-linear load such as electric arc furnaces, variable speed drives, large concentrations of arc discharge lamps, saturation of magnetization of transformer and a distorted line current. The current generated by such load interact with power system impedance and gives rise to harmonics. The effect of harmonics in the power system can lead to degradation of power quality at the consumer's terminal, increase of power losses, and malfunction in communication system. The degree of variation is assessed at the point of common connection, where consumer and supplier area of responsibility meet. The harmonics voltage and current should be limited to acceptable level at the point of wind turbine connection in the system. This fact has lead to more stringent requirements regarding power quality, such as Standard IEC 61000-3-2 or IEEE-519. Conventionally, passive LC resonant filters have been used to solve power quality problems. However, these filters have the demerits of fixed compensation, large size, and the resonance itself. To overcome these drawbacks, active filters appear as the dynamic solution.

The IEC 61000-3-6 gives a guideline and harmonic current limits. According to standard IEC 61400-21 guideline, harmonic measurements are not required for fixed speed wind turbines where the induction generator is directly connected to grid. Harmonic measurements are required only for variable speed turbines equipped with electronic power converters. In general the power converters of wind turbines are pulse-width modulated inverters, which have carrier frequencies in the range of 2-3 kHz and produce mainly inter harmonic currents.

The harmonic measurement at the wind turbine is problem due to the influence of the already existing harmonic voltage in the grid. The wave shape of the grid voltage is not sinusoidal. There are always harmonics voltages in the grid such as integer harmonic of 5th and 7th order which affect the measurements.

Today's variable speed turbines are equipped with self commutated PWM inverter system. This type of inverter system has advantage that both the active and reactive power can be controlled, but it also produced a harmonic current. Therefore filters are necessary to reduce the harmonics.

The harmonic distortion is assessed for variable speed turbine with a electronic power converter at the point of common connection. The total harmonic voltage distortion of voltage is given as in (4).

$$V_{THD} = \sqrt{\sum_{h=2}^{40} \frac{V_h^2}{V_1} 100} \tag{4}$$

V_h- h[th] harmonic voltage and V_1 –fundamental frequency 50 Hz. The THD limit for various level of system voltages are given in the table 1.0

System Voltage (kV)	Total Harmonic Distortion (%)
400	2.0
220	2.5
132	3.0

Table 1. Voltage Harmonics Limit

THD of current I_{THD} is give as in (5)

$$I_{THD} = \sqrt{\sum_{h=2}^{40} \frac{I_h^2}{I_1} 100} \tag{5}$$

Where I_h - h[th] harmonic current and I_1 –fundamental frequency (50) Hz. The acceptable level of THD in the current is given in table 2.

Voltage level	66 kV	132kV
I_{THD}	5.0	2.5

Table 2. Current Harmonic Limit

Various standards are also recommended for individual consumer and utility system for helping to design the system to improve the power quality. The characteristics of the load and level of power system significantly decides the effects of harmonics. IEEE standards are adapted in most of the countries. The recommended practice helps designer to limit current and voltage distortion to acceptable limits at point of common coupling (PCC) between supply and the consumer.

1. IEEE standard 519 issued in 1981, recommends voltage distortion less than 5% on power lines below 69 kV. Lower voltage harmonic levels are recommended on higher supply voltage lines.

2. IEEE standard 519 was revised in 1992, and impose 5% voltage distortion limit. The standards also give guidelines on notch depth and telephone interface considerations.
3. ANSI/IEEE Standard C57.12.00 and C57.12.01 limits the current distortion to 5% at full load in supply transformer.

In order to keep power quality under limit to a standards it is necessary to include some of the compensator. Modern solutions for active power factor correction can be found in the forms of active rectification (active wave shaping) or active filtering.

2.5. Flickers

Flicker is the one of the important power quality aspects in wind turbine generating system. Flicker has widely been considered as a serious drawback and may limit for the maximum amount of wind power generation that can be connected to the grid. Flicker is induced by voltage fluctuations, which are caused by load flow changes in the grid. The flicker emission produced by grid-connected variable-speed wind turbines with full-scale back-to-back converters during continuous operation and mainly caused by fluctuations in the output power due to wind speed variations, the wind shear, and the tower shadow effects. The wind shear and the tower shadow effects are normally referred to as the 3p oscillations. As a consequence, an output power drop will appear three times per revolution for a three-bladed wind turbine. There are many factors that affect flicker emission of grid connected wind turbines during continuous operation, such as wind characteristics and grid conditions. Variable-speed wind turbines have shown better performance related to flicker emission in comparison with fixed-speed wind turbines.

The flicker study becomes necessary and important as the wind power penetration level increases quickly. The main reason for the flicker in fixed speed turbines is to wake of the tower. Each time a rotor blade passes the tower, the power output of the turbine is reduced. This effect cause periodical power fluctuations with a frequency of about ~1 Hz. The power fluctuation due to the wind speed fluctuation has lower frequencies and thus is less critical for flicker. In general, the flicker of fixed speed turbines reaches its maximum at high wind speed. Owing to smoothing effect, large wind turbine produced lower flicker than small wind turbines, in relation to their size.

Several solutions have been proposed to mitigate the flicker caused by grid-connected wind turbines. The mostly adopted technique is the reactive power compensation. It can be realized by the grid-side converter of variable-speed wind turbines or the Static synchronous compensator connected at the point of common coupling (PCC). Also, some papers focus on the use of active power curtailment to mitigate the flicker [5].

The flicker level depends on the amplitude, shape and repetition frequency of the fluctuated voltage waveform. Evaluating the flicker level is based on the flicker meter described in IEC 61000-4-15. Two indices are typically used as a scale for flicker emission, short-term flicker index, P_{st} and long-term flicker index, P_{lt}. P_{lt} is estimated by certain process of the P_{st} values.

It is assumed that wind turbines under study is running at normal operation; hence, the long-term flicker index (P_{lt}), which is based on a 120-min time interval, is equal to P_{st} and, therefore, P_{st} is only considered in this work. The normalized response of the flicker meter described in Figure 2.0.

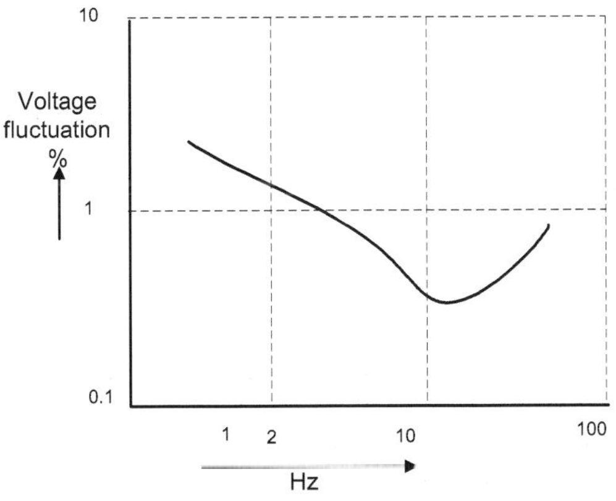

Figure 2. Influence of frequency on the perceptibility of sinusoidal voltage change

A quite small voltage fluctuation at certain frequency (8.8 Hz) can be irritable. The flicker level ($P_{st} \leq 1$) is a threshold level for connecting wind turbines to low voltage. The measurements are made for maximum number of specified switching operation of wind turbine with 10-minutes period and 2-hour period are specified, as given in (6)

$$P_{lt} = C(\Psi_K)\frac{S_n}{S_K} \tag{6}$$

Where p_{lt} - Long term flicker. $C(\Psi_K)$ - Flicker coefficient calculated from Rayleigh distribution of the wind speed. The Limiting Value for flicker coefficient is about ≤ 0.4, for average time of 2 hours.

2.6. Reactive power

Traditional wind turbines are equipped with induction generators. Induction generator is preferred because they are inexpensive, rugged and requires little maintenance. Unfortunately induction generators require reactive power from the grid to operate. The interactions between wind turbine and power system network are important aspect of wind generation system. When wind turbine is equipped with an induction generator and fixed capacitor are used for reactive compensation then the risk of self excitation may occur during off grid operation. Thus the sensitive equipments may be subjected to over/under voltage, over/under frequency operation and other disadvantage of safety aspect. According to IEC Standard, reactive power of wind turbine is to be specified as 10 min average value as a function of 10-min. output power for 10%, 20% ... 100% of rated power. The effective control of reactive power can improve the power quality and stabilize the grid. Although reactive power is unable to provide actual working benefit, it is often used to adjust voltage, so it is a useful tool for maintaining desired voltage level. Every transmission system always has a reactive component, which can be expressed as power factor. Thus the some method is needed to manage the reactive power by injecting or absorbing VAr as necessary in order to maintain optimum voltage level and enable real power flow. Until recently, this has been especially difficult to effectively accomplish at a wind farms due to the variable nature of wind. The suggested control technique in the thesis is capable of controlling reactive power to zero value at point of common connection (PCC).The mode of operation is referred as unity power factor.

2.7. Location of wind turbine

The way of connecting wind turbine into the electric power system highly influences the impact of the wind turbine generating system on the power quality. As a rule, the impact on power quality at the consumer's terminal for the wind turbine generating system (WTGS) located close to the load is higher than WTGS connected away, that is connected to H.V. or EHV system.

Wind turbine generator systems (WTGS) are often located in the regions that have favorable wind conditions and where their location is not burdensome. These regions are low urbanized, which means that the distribution network in these regions is usually weak developed. Such situation is typical for all countries developing a wind power industry.

The point of common coupling (PCC) of the WTGS and the power network parameter and structure of grid is of essential significance in the operation of WTGS and its influence on the system. WTGS can be connected to MV transmission line and to HV networks.

The WTGS connected to the existing MV transmission line, which feeds the existing customers is presented in Figure 3.

The distance between WTGS and PCC is usually small up to a few kilometers. Such connections are cheap as compare to other types of connection but greatly affected on consumers load (power quality).

If the location of WTGS is connected to an MV bus in feeding an HV/MV substation through a separate transmission line (position 1), the connection has some advantages related to low influence of WTGS on customers load. Such connection are expensive than presented above.

The location of WTGS connected to HV bus through a separate transmission line, when a relatively large rated WTGS has to be connected in the power network, where the MV network is weak. This type of connection are most expensive than other presented.

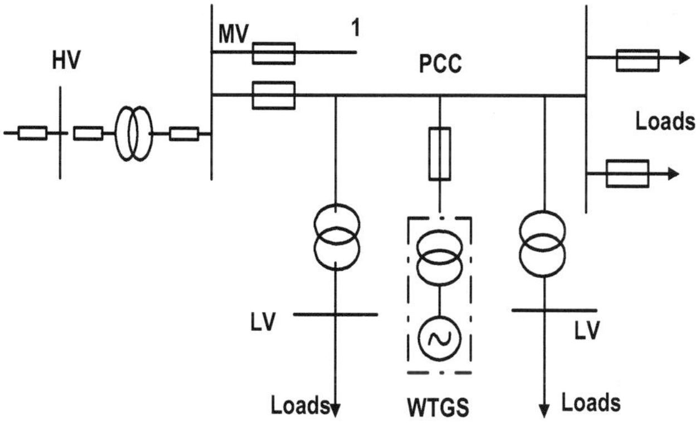

Figure 3. WTGS coupled to MV transmission-line.

2.8. Low voltage ride through capability

The impact of the wind generation on the power system will no longer be negligible if high penetration levels are going to be reached. The extent to which wind power can be integrated into the power system without affecting the overall stable operation depends on the technology available to mitigate the possible negative impacts such as loss of generation for frequency support, voltage flicker, voltage and power variation due to the variable speed

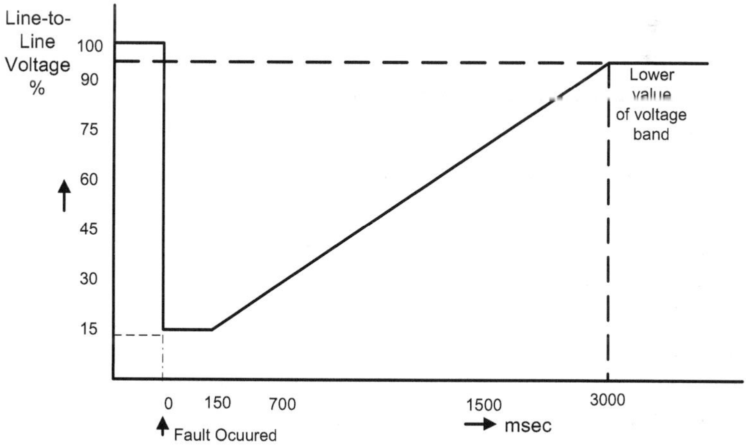

Figure 4. Low voltage ride through (LVRT) capability

of the wind and the risk of instability due to lower degree of controllability. Many countries in Europe and other parts of the world are developing or modifying interconnection rules and processes for wind power through a grid code. The grid codes have identified many potential adverse impacts of large scale integration of wind resources. The risk of voltage collapse for lack of reactive power support is one of the critical issues when it comes to contingencies in the power system. The low voltage ride through (LVRT) capability, which is one of the most demanding requirement that have been included in the grid codes and shown in Fig. 4.

It defines the operational boundary of a wind turbine connected to the network in terms of frequency, voltage tolerance, power factor, fault ride through is regarded as the main challenges to the wind turbine manufactures. The wind turbine should remain stable and connected during the fault while voltage at the PCC drop to 15% of the nominal value i.e. drops of 85% for the part of 150 msec. Only when the grid voltage fall below the curve, the turbine is allowed to disconnected from the grid.

Significant barriers to interconnection are being perceived already with the requirements of the new grid codes and there it is a need for a better understanding of the factors affecting the behavior of the wind farm under severe contingencies such as voltage sags. Wind farms using squirrel cage induction generators directly connected to the network will suffer from the new demands, since they have no direct electrical control of torque or speed, and would usually disconnect from the power system when the voltage drops more than 10–20% below the rated value. In general, fulfillment of LVRT by reactive compensation will require fast control strategies for reactive power in wind turbines/farms with cage induction generators. The LVRT requirement, although details are differing from country to country, basically demands that the wind farm remains connected to the grid for voltage dips as low as 5%.

2.9. IEC recommendation

For consistent and replicable documentation of power quality characteristic of wind turbine, the international Electro-technical Commission IEC-61400-21 was developed and today, most of the large wind turbine manufactures provide power quality characteristic data accordingly.IEC 61400-21 describe the procedures for determine the power quality characteristics of wind turbines. It is a guideline for power quality measurements of wind turbine. The methodology of IEC standard consists of three analyses. The first one is the flickers analyses. IEC 61400-21 specified a method that uses current and voltage time series measured at the wind turbine terminals to simulate the voltage fluctuation on a fictitious grid with no source of voltage fluctuations other that wind turbine switching operation. The second one is regarding the switching operation. The voltage and current transients are measured during the switching operation of wind turbine. The last one is the harmonic analysis which is carried out by FFT algorithms. Recently harmonic and inter harmonic are treated in the IEC 61000-4-7 and IEC 61000-3-6. The method for summing harmonics and inter harmonic in the IEC 61000-3-6 are applicable to wind turbines. The inter harmonics that are not a multiple of 50 Hz, since the switching frequency of the inverter is not constant but varies, the harmonic will also vary. Consequently, the grid codes has been define to specify the requirements that the wind turbines must meet in order to be connected to the grid, including the capabilities of contributing to frequency and voltage control by adjusting the active and reactive power supplied to the transmission system.

3. Grid code for wind farms

The Electricity Grid Code is a regulation made by the Central Commission and it to be follow by various persons and participants in the system to plan, develop, maintain, and operate the power system grid in the most secure, reliable, economic and efficient manner, while facilitating healthy competition in the generation and supply of electricity.

The first grid code was focused on the distribution level, after the blackout in the United State in August 2003. The United State wind energy industry took a stand in developing its own grid code for contributing to a stable grid operation. The rules for realization of grid operation of wind generating system at the distribution network is defined as - per IEC-61400-21.The grid quality characteristics and limits are given for references that the customer and the utility grid may expect. According to Energy-Economic Law, the operator of transmission grid is responsible for the organization and operation of interconnected system. The grid code also covers some of the technical standards for connection to the grid.

To ensure the safe operation, integrity and reliability of the grid is utmost important. It is mentioned that reactive power compensation should ideally be provided locally by generating reactive power as close to the reactive power consumption as possible. The regional entity except generating stations, expected to provide local VAr compensation/generation such that they do not draw VArs from the grid, particularly under low-voltage condition. Indian grid code commission mentions that the charge for VArh shall be at the rate of 25 paise/kVArh w.e.f.1.4.2010, for VAr interchanges.

The wind farms must be able to run at rated voltage at a specified voltage range. The voltage range depends on the level of the voltage on the transmission system, which varies from country to country.

The wind farms shall have a closed loop voltage regulation system. The voltage regulation system shall act to regulate the voltage at the point by continuous modulation of the reactive power output within its reactive power range, and without violating the voltage step emissions.

Voltage fluctuations at a point of common coupling with a fluctuating load directly connected to the transmission system shall not exceed 3% at any time. The flicker contributions P_{st} and P_{lt} are defined in IEC 61000-3-7 (Electromagnetic compatibility).

The wind turbine generator (WTG) shall be equipped with voltage and frequency relays for disconnection of the wind farm at abnormal voltages and frequencies. The relays shall be set according to agreements with the regional grid company and the system operator. Following are the technical requirements to be fulfilled to integrate the wind generation system.

Voltage Rise (u) -The voltage rise at the point of common coupling can be approximated as a function of maximum apparent power S_{max} of the turbine, the grid impedances R and X at the point of common coupling and the phase angle ϕ. The Limiting voltage rise value is < 2%

Voltage dips (d) - The voltage dips is due to start up of wind turbine and it causes a sudden reduction of voltage. The acceptable voltage dips limiting value is ≤ 3 %.

Flicker-The measurements are made for maximum number of specified switching operation of wind turbine with 10-minutes period and 2-hour period are specified. The Limiting Value for flicker coefficient is about ≤ 0.4, for average time of 2 hours

$$Harmonics - \ The\ THD\ limit\ for\ 132\ KV\ is\ <\ 3\%.$$

Grid frequency- The grid frequency in India is specified in the range of 47.5-51.5 Hz, for wind farm connection. The wind farm shall able to withstand change in frequency up to 0.5Hz/sec. Thus the requirements in the Grid Code can be fulfill the technical limits of the network.

4. Conclusion

The chapter provides the challenges regarding the integration of wind energy in to the power systems. Today the worldwide trend of wind power penetration is increased. The integration of high penetration level of wind power into existing power system has significant impact on the power system operation.

The wind turbines connected to weak grids have an important influence on power system. The weak grid is characterized by large voltage and frequency variations, which affects wind turbines regarding their power performance, safety and allied electrical components. The strength of the distribution system is important from the point of power quality. The needs for consistent qualification of power quality characteristics of wind turbines, the International Electro-Technical Commission to facilitate for power quality parameters for various issues are presented. The latest grid code requirements are to ensure that wind farms do not adversely affect the power system operation with respect to security of supply, reliability and for power quality.

Author details

Sharad W. Mohod
Ram Meghe Institute of Technology & Research,Badnera-Amravart, India

Mohan V. Aware
Visvesvaraya National Institute of Technology, Nagpur, India

5. References

[1] H. Holitinen, R.Hirvonen, "Power system requirement for wind power.", Wind Power in Power System, T. Ackermann, Ed. New-York, pp 143-157, Wiley, 2005

[2] R.C. Bansal, Ahmed F. Zobaa, R.K. Saket, "Some issue related to power generation using wind energy conversion system: An overview" *Int. Journal of Emerging Electric Power System*, Vol. 3, No.2, pp 1-14, 2005.

[3] J.Charles Smith, Michael R. Milligan, Edgar A. DeMeo, "Utility wind integration and operating impact state of the art., *IEEE Trans on Energy Conversion* Vol.22, No.3, pp 900-907, August 2007.

[4] Z.Sadd-Saoud, N. Jenkins, "Models for predicting flicker induced by large wind turbines" *IEEE Trans on Energy Conversion*, Vol.14, No.3, pp.743-751, Sept.1999.

[5] Weihao Hu, Zhe Chen, Yue Wang and Zhan Wang "Flicker Mitigation by Active Power Control of Variable-Speed Wind Turbines With Full-Scale Back-to-Back Power Converters", *IEEE Trans on Energy Conversion*, Vol. 24, No. 3, pp.640-648, Sept. 2009

[6] F.Zhou, G.Joos, C.Abhey. "Voltage stability in weak connection wind farm." *IEEE PES Gen. Meeting*, Vol. 2, pp.1483-1488, 2005

[7] S.W.Mohod, M.V.Aware, "Power quality issues & its mitigation technique in wind generation, "*Proc. of IEEE Int. Conf. on Harmonics and Quality of Power* (ICHQP), pp. 1-6, Sept. 2008

[8] S.Z. Djokic, J.V. Milanovic, "Power quality and compatibility levels: A general approach.", *IEEE Trans on Power Delivery*, Vol.22, No.3, pp.1857-1862, July. 2007

[9] Helder J. Azevedo, Jose M. Ferreiraa, Antonio P. Martins, Adriano S. Carvalho, "An active power filter with direct current control for power quality conditioning." *Electric Power Component and System*, Vol.26, pp.587-601, 2008.

[10] Z.Chen, E. Spooner, "Grid power quality with variable speed wind turbines", *IEEE Trans on Energy Conversion*, Vol.16, No.2, pp.148-156, June 2001.

[11] Dusan Graovac, Vladimir A. Katic, Alfred Rufer, "Power quality problems compensation with universal power quality conditioning system." *IEEE Trans on Power Delivery*, Vol.22, No.2, pp.968-975, April 2007.

[12] Juan Manuel Carrasco, Leopoldo Garcia Franquelo, Jan. Bialasiewicz, "Power electronic system for the grid integration of renewable energy sources: A survey." *IEEE Trans on Industry Electronics*, Vol.53, No.4, pp.1002-1010, Aug. 2006.

[13] "A report on Indian Wind Grid Code" – *Committee draft*, Version 1.0, pp 5-42 July 2009.

[14] J.F.Conroy, R. Watson, "Low-voltage ride though of full converter wind turbine with permanent magnet generator", *Proc. IET Renew. Power Generation.* Vol.2, No.2, pp.113-122, 2008.

[15] Bousseau, P.: "Solution for the grid integration of wind farms – A survey", *Proc. of European Wind Energy Conf.*, pp.12-16, Nov. 2004.

[16] S.W.Mohod, M.V.Aware, "A STATCOM-Control Scheme for Grid Connected Wind Energy System for Power Quality Improvement", *IEEE System Journal*, A special issue on sustainable development, Vol.4, Issue-3, pp-346-352, Sept-2010.

[17] S.W.Mohod, S.M.Hatwar, M.V.Aware. "Wind Energy Generation Interfaced System with Power Quality and Grid Support" Journal of Advanced Materials Research, Volumes 403 - 408, Tech Publication Swizerland, pp.2079-2086, Nov.2011.

Section 2

Power Quality Improvement in Transmission and Distribution Systems

Chapter 3

Impact of Series FACTS Devices (GCSC, TCSC and TCSR) on Distance Protection Setting Zones in 400 kV Transmission Line

Mohamed Zellagui and Abdelaziz Chaghi

Additional information is available at the end of the chapter

http://dx.doi.org/10.5772/54257

1. Introduction

The electricity supply industry is undergoing a profound transformation worldwide. Market forces, scarcer natural resources, and an ever-increasing demand for electricity are some of the drivers responsible for such unprecedented change. Against this background of rapid evolution, the expansion programs of many utilities are being thwarted by a variety of well-founded, environment, land-use, and regulatory pressures that prevent the licensing and building of new transmission lines and electricity generating plants.

The ability of the transmission system to transmit power becomes impaired by one or more of the following steady state and dynamic limitations:

- Angular stability,
- Voltage magnitude,
- Thermal limits,
- Transient stability,
- Dynamic stability.

These limits define the maximum electrical power to be transmitted without causing damage to transmission lines and electrical equipment. In principle, limitations on power transfer can always be relieved by the addition of new transmission lines and generation facilities.

Alternatively, Flexible Alternating Current Transmission System (FACTS) controllers can enable the same objectives to be met with no major alterations to power system layout. FACTS are alternating current transmission systems incorporating power electronic-based and other static controllers to enhance controllability and increase power transfer capability.

The FACTS concept is based on the substantial incorporation of power electronic devices and methods into the high-voltage side of the network, to make it electronically controllable.

FACTS controllers aim at increasing the control of power flows in the high-voltage side of the network during both steady state and transient conditions. Owing to many economical and technical benefits it promised, FACTS received the support of electrical equipment manufacturers, utilities, and research organizations around the world. This interest has led to significant technological developments of FACTS controllers (Sen, K.K.; Sen, M.L., 2009), (Zhang, X.P. et al., 2006). Several kinds of FACTS controllers have been commissioned in various parts of the world.

Popular are: load tap changers, phase-angle regulators, static VAR compensators, thyristors controlled series compensators, interphase power controllers, static compensators, and unified power flow controllers.

The main objectives of FACTS controllers are the following (Mathur, R.M.; Basati, R.S., 2002):

- Regulation of power flows in prescribed transmission routes,
- Secure loading of transmission lines nearer to their thermal limits,
- Prevention of cascading outages by contributing to emergency control,
- Damping of oscillations that can threaten security or limit the usable line capacity.

The most Utility engineers and consultants use relay models to select the relay types suited for a particular application, and to analyze the performance of relays that appear to either operate incorrectly or fail to operate on the occurrence of a fault. Instead of using actual prototypes, manufacturers use relay model designing to expedite and economize the process of developing new relays. Electric power utilities use computer-based relay models to confirm how the relay would perform during systems disturbances and normal operating conditions and to make the necessary corrective adjustment on the relay settings. The software models could be used for training young and inexperienced engineers and technicians. Researchers use relay model to investigate and improve protection design and algorithms. However, simulating numerical relays to choose appropriate settings for the steady state operation of over current relays and distance relays is presently the most familiar use of relay models (McLaren et al., 2001).

1.1. Problem statement

In the presence of series compensators the system FACTS devices i.e. GTO Controlled Series Capacitor (GCSC), Thyristor Controlled Series Capacitor (TCSC) and Thyristor Controlled Series Reactor (TCSR) connected in high voltage (HV) transmission line protected by distance relay, the total impedance and the measured impedance at the relaying point depend on the injected reactance by compensators. So there is a reel impact on the relay settings zones.

1.2. Objectives

This chapter presents a comparative study of the performance of MHO (admittance) distance relays for transmission line 400 kV in Eastern Algerian transmission networks compensated by three different series FACTS i.e. GCSC, TCSC and TCSR connected at midpoint of a single electrical transmission line. The facts are used for controlling transmission voltage in the range of ±40kV as well as reactive power injected between -50 MVar/+15 MVar on the power system. This chapter studies the effects of GCSC, TCSC and TCSR insertion on the total impedance of a transmission line protected by MHO (admittance) distance relay.

The modified setting zone protection in capacitive and inductive boost mode for three forward zones (Z_1, Z_2 and Z_3) and reverse zone (Z_4) have been investigated in order to prevent circuit breaker nuisance tripping to improve the performances of distance relay protection. The simulation results are performed in MATLAB software.

2. Apparent reactance injected by series FACTS devices

In general, FACTS compensator can be divided into three categories (Acha, E. al., 2004): Series compensator, Shunt compensator, and combined series-series compensator. In this chapter, we study the series FACTS devices.

2.1. GCSC

The compensator GCSC mounted on figure 1.a is the first that appears in the family of series compensators. It consists of a capacitance (C) connected in series with the transmission line and controlled by a valve-type GTO thyristors mounted in anti-parallel and controlled by an angle of extinction (γ) varied between 0° and 180°. If the GTOs are kept turned-on all the time, the capacitor C is bypassed and it does not realize any compensation effect. On the other hand, if the positive-GTO (GTO_1) and the negative-GTO (GTO_2) turn off once per cycle, at a given angle γ counted from the zero-crossing of the line current, the main capacitor C charges and discharges with alternate polarity (Zhang, X.P. et al., 2006), (De Jesus F. D. et al., 2007).

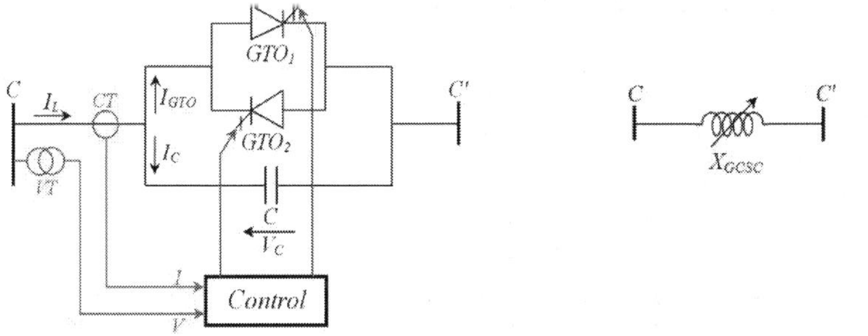

(a). Mounting. (b). Apparent reactance.

Figure 1. Transmission line in presence of GCSC

Hence, a voltage V_C appears in series with the transmission line, which has a controllable fundamental component that is orthogonal (lagging) to the line current.

The compensator GCSC injects in the transmission line a variable capacitive reactance (X_{GCSC}). From figure 1.b the expression of X_{GCSC} is directly related to the controlled GTO angle (γ) which is varied between 0° and 180° as expressed by following equation (De Souza, L. F. W. et al., 2008), (Ray, S. et al., 2008) :

$$X_{GCSC}(\gamma) = X_{C\max}\left[1 - \frac{2}{\pi}\gamma - \frac{1}{\pi}\sin(2\pi)\right] \quad (1)$$

Where,

$$X_{C\max} = 1/C.\omega \quad (2)$$

2.2. TCSC

The compensator TCSC mounted on Figure 2.a is a type of series FACTS compensators. It consists of a capacitance (C) connected in parallel with an inductance (*L*) controlled by a valve mounted in anti-parallel conventional thyristors (T_1 and T_2) and controlled by an angle of extinction (α) varied between 90° and 180°.

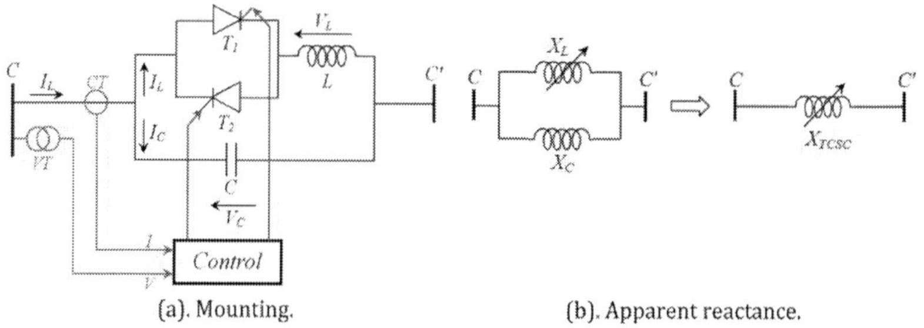

Figure 2. Transmission line in presence of TCSC

From figure 2.b, the compensator TCSC injected in the transmission line a variable capacitive reactance (X_{TCSC}). The expression of X_{TCSC} is directly related to the controlled thyristors, angle (α) which is varied between 90° and 180° and expressed by following equation (Acha, E. al., 2004), (Sen, K.K.; Sen, M.L., 2009):

$$X_{TCSC}(\alpha) = X_C // X_L(\alpha) = \frac{X_C.X_L(\alpha)}{X_C + X_L(\alpha)} \quad (3)$$

$$X_L(\alpha) = X_{L\max}\left[\frac{\pi}{\pi - 2\alpha - \sin(2\alpha)}\right] \quad (4)$$

Where,

$$X_{L\max} = L.\omega \quad (5)$$

And,

$$X_C = {}^{-1}\!/_{j.C.\omega} \tag{6}$$

From the equations (4), (5) and (6), the equation (3) becomes:

$$X_{TCSC}(\alpha) = \frac{X_C.X_{L.\max}\left[\dfrac{\pi}{\pi - 2\alpha - \sin(2\alpha)}\right]}{X_C + X_{L.\max}\left[\dfrac{\pi}{\pi - 2\alpha - \sin(2\alpha)}\right]} \tag{7}$$

2.3. TCSR

The compensator TCSR is an inductive reactance compensator at which its inductive reactance is continually adjusted through the firing delay angle (α) of the thyristors as shown in figure 3.a. It consists of a series reactor shunted by a thyristors controlled reactor (TCR).

If the firing delay angle is 180°, the TCSR operates as an uncontrolled reactor (L_1). When the angle decreases below 180°, the inductive reactance of TCSR decreases and at 90° it is given by the parallel connection of the reactors ($L_1//L_2$).

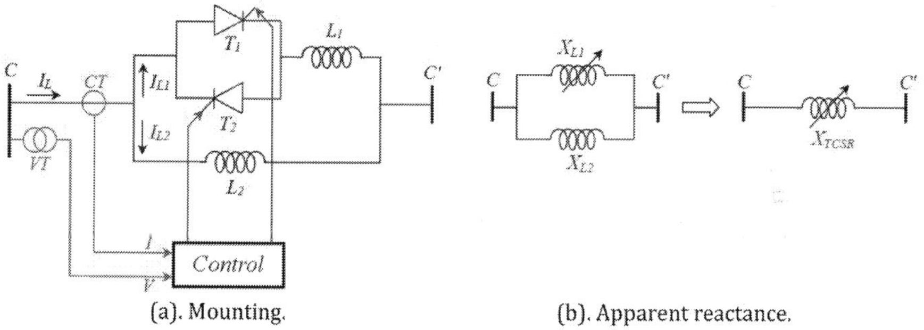

(a). Mounting. (b). Apparent reactance.

Figure 3. Transmission line in presence of TCSR.

From figure 3.b, the compensator TCSR injected in the transmission line a variable capacitive reactance (X_{TCSR}). The expression of XTCSR is directly related to the controlled thyristors angle (α) expressed by the following equation (Acha, E. al., 2004), (Zhang, X.P. et al., 2006):

$$X_{TCSR}(\alpha) = X_{L2}//X_{L1}(\alpha) = \frac{X_{L2}.X_{L1}(\alpha)}{X_{L2} + X_{L1}(\alpha)} = \frac{X_{L2}.X_{L1-\max}\left[\dfrac{\pi}{\pi - 2\alpha - \sin(2\alpha)}\right]}{X_{L2} + X_{L1-\max}\left[\dfrac{\pi}{\pi - 2\alpha - \sin(2\alpha)}\right]} \tag{8}$$

Where,

$$X_{L1-\max} = L_1.\omega \tag{9}$$

And,

$$X_{L2} = j.L_2.\omega \tag{10}$$

3. Power system protection

Fault current is the expression given to the current that flow in the circuit when load is shorted i.e. flow in a path other than the load. This current is usually very high and may exceed ten times the rated current of a piece of plant. Faults on power system are inevitable due to external or internal causes, lightning may struck the overhead lines causes insulation damage. Internal overvoltage due to switching or other power system phenomenon may also cause an over voltage which leads to deterioration of the insulation and faults. Power networks are usually protected by means of two main components, relays that sense the abnormal current or voltage and a circuit breaker that put a piece of plant out of tension.

Power system protection is the art and science of the application of devices that monitor the power line currents and voltages (relays) and generate signals to deenergize faulted sections of the power network by circuit breakers. Goal is to minimize damage to equipment that would be caused by system faults, if residues, and maintain the delivery of electrical energy to the consumers (Horowitz, S.H.; Phadke A.G. 2008), (Blackburn, J.L.; Domin, T.J. 2006).

Many types of protective relays are used to protect power system equipments. They are classified according to their operating principles; over current relay senses the extra (more than set) current considered dangerous to a given equipment, differential relays compare in and out currents of a protected equipment, while impedance relays measure the impedance of the protected piece of plant.

3.1. Principal characteristics of protection system

For system protection to be effective, the following characteristics must be met (Blackburn J.L.; Domin. T.J., 2006), (Zellagui. M, Chaghi. A., 2012):

- Reliability: assurance that the protection will perform correctly in presence of faults on electrical transmission and distribution line,
- Selectivity: maximum continuity of service with minimum system disconnection,
- Speed of operation: minimum fault duration and consequent equipment damage and system instability,
- Simplicity: minimum protective equipment and associated circuitry to achieve the protection objectives,
- Economics: maximum protection at minimal total cost.

3.2. Principles of relay application

The power system is divided into protection zones defined by the equipment and the available circuit breakers. Six categories of protection zones are possible in each power system:

- Generators and generator-transformer units,
- Transformers,
- Bus bars,
- Lines (transmission and distribution),
- Utilization equipment (motors, static loads, or other),
- Capacitor or reactor banks (when separately protected).

3.3. Protection zones

Most of these zones are illustrated in figure 4. Although the fundamentals of protection are quite similar, each of these six categories has protective relays, specifically designed for primary protection, that are based on the characteristics of the equipment being protected. The protect ion of each zone normally include s relays that can provide backup for the relays protecting the adjacent equipment (Zellagui.M; Chaghi.A. 2012.a). The protection in each zone should overlap that in the adjacent zone; otherwise, a primary protection void would occur between the protection zones. This overlap is accomplished by the location of the CTs the key sources of power system information for the relays.

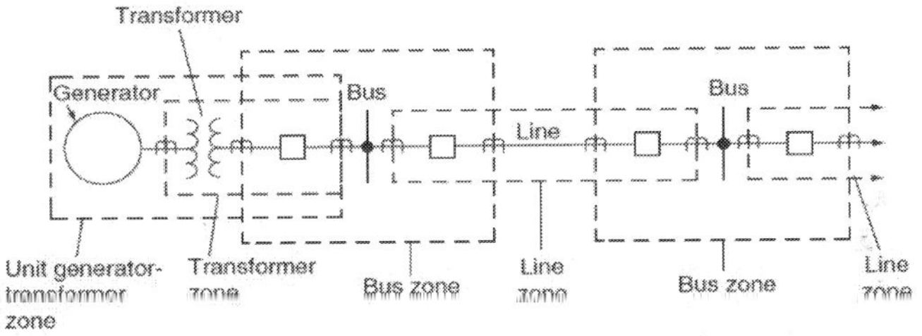

(a). Typical relay for generator, line and bus.

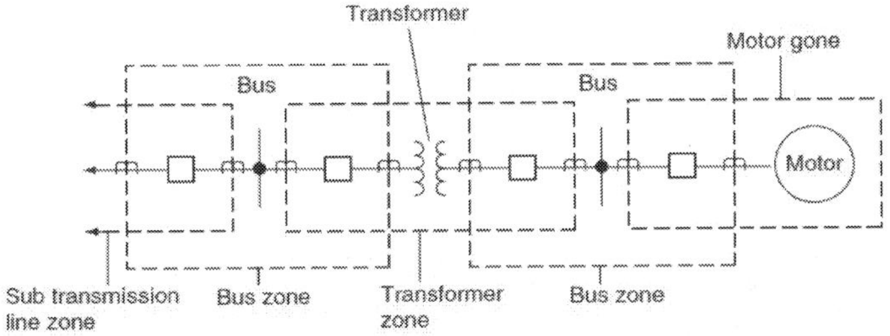

(b). Typical relay for motor and transformer.

Figure 4. Protection zone on power system

4. Setting zones for MHO distance relays

4.1. Principal

Distance protection is so called because it is based on an electrical measure of distance along a transmission line to a fault. The distance along the transmission line is directly proportional to the series electrical impedance of the transmission line.

Impedance is defined as the ratio of voltage to current. Therefore, distance protection measures distance to a fault by means of a measured voltage to measured current ratio computation (Zigler, G., 2008), (Zellagui, M.; Chaghi, A., 2012.b). The philosophy of setting relay at Sonelgaz Group is three forward zones and one reverse zone to protect EHV transmission line between busbar *A* and *B* with total impedance Z_{AB} as shown in figure 5.

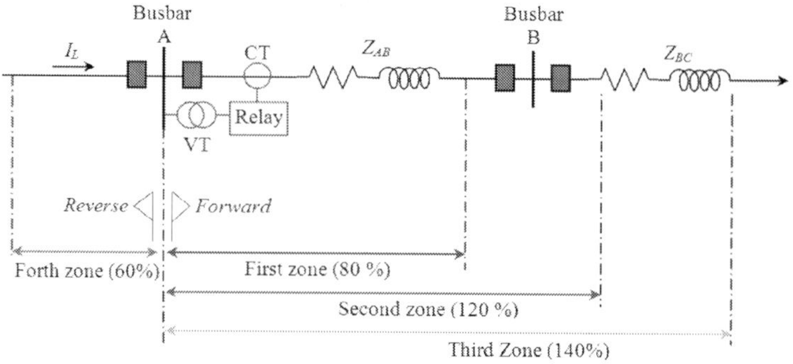

Figure 5. Principal operation of distance relay

4.2. Setting zones

4.2.1. First zone

In practice it is normal to adjust the first zone relays (Z_1) at A to protect only up to 80% of the protective line AB. This is a high speed unit and is used for the primary protection of the protected line. Its operation is instantaneous (Dechphung, S.; Saengsuwan, T., 2008).

This unit is not set to protect the entire line to avoid undesired tripping due to over reach. Over reach may occur due to transients during the fault condition.

4.2.2. Second zone

It is set to cover about 20% of the second line (BC). The main object of the second zone unit is to provide protection to the end zone of the first section which is beyond the reach of the first unit. The setting of the second unit is so adjusted that it operates the relay even for arcing faults at the end of the line. To achieve this, the unit must take care beyond the end of the line. In other words its setting must take care of under reach caused by arc resistance (Dechphung, S; Saengsuwan, T., 2008), (Zellagui, M.; Chaghi, A., 2012.b).

Under reach is also caused by intermediate current sources, errors in CT, and VT and measurement performed by the relay. To take into account the under reaching tendency caused by these factors, the normal practice is to set the second zone reach up to 20% of the shortest adjoining line section. The protective zone of the second unit is known as the second zone of protection. The second zone unit operates after a certain time delay. Its operating time is 0,3 sec.

4.2.3. Third zone

It is provided for back-up protection of the adjoining line. Its reach should extend beyond the end of the adjoining line under the maximum under reach, which may be caused by arcs, intermediate current sources and errors in CT, VT and measuring unit (Zellagui. M.; Chaghi. A., 2012.b). The protective zone of the third stage is known as the third zone of protection.

The characteristic curve on MHO (admittance) relay for setting zones is shown in figure 6.

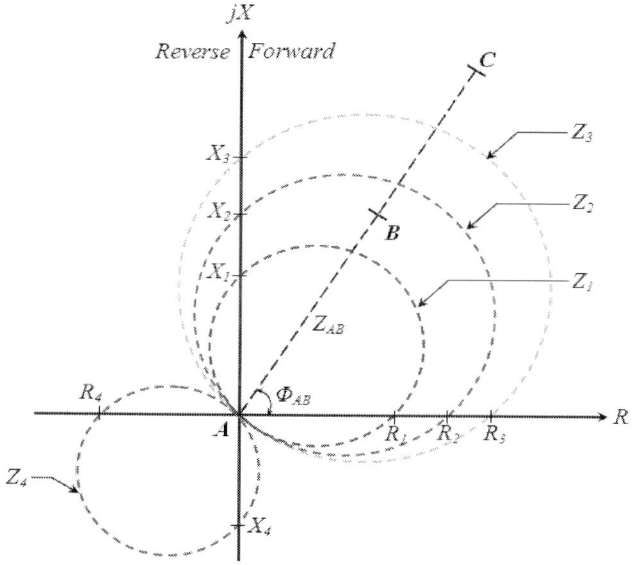

Figure 6. Characteristic curve X (R) for setting zones for distance protection.

Figure 7 represents the tripping time T_1, T_2 and T_3 correspond to these three zones of operation for circuit breaker installed at busbar A and MHO distance relay (R_A).

The fourth setting zones for protected transmission line (forward and reverse) without series FACTS are given by (Zellagui, M.; Chaghi, A. 2012.c), (Gérin-Lajoie, L. 2009):

$$Z_1 = R_1 + jX_1 = 80\% Z_{AB} = 0{,}8.(R_{AB} + jX_{AB}) \quad (11)$$

$$Z_2 = R_2 + jX_2 = R_{AB} + jX_{AB} + 0,2.(R_{BC} + jX_{BC}) \quad (12)$$

$$Z_3 = R_3 + jX_3 = R_{AB} + jX_{AB} + 0,4.(R_{BC} + jX_{BC}) \quad (13)$$

$$Z_4 = R_4 + jX_4 = -60\% Z_{AB} = -0,6.(R_{AB} + jX_{AB}) \quad (14)$$

The total impedance of transmission line AB measured by MHO distance relay is:

$$Z_{AB} = K_Z.Z_L, \quad K_Z = {K_{VT}}/{K_{CT}} \quad (15)$$

Where, ZAB is real total impedance of line AB, and KVT and KCT is ratio of voltage to current respectively.

The presence of series FACTS systems in a reactor (X_{FACTS}) has a direct influence on the total impedance of the protected line (Z_{AB}), especially on the reactance XAB and no influence on the resistance RAB.

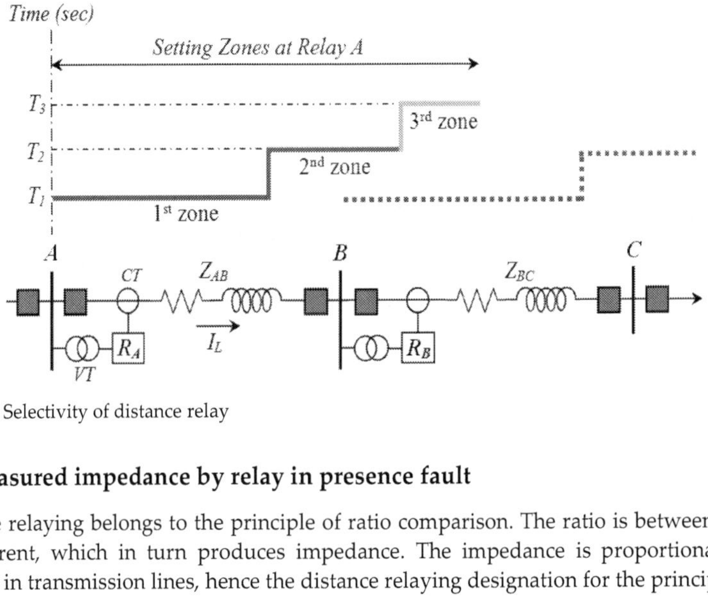

Figure 7. Selectivity of distance relay

4.3. Measured impedance by relay in presence fault

Distance relaying belongs to the principle of ratio comparison. The ratio is between voltage and current, which in turn produces impedance. The impedance is proportional to the distance in transmission lines, hence the distance relaying designation for the principle.

This principle is primarily used for protection of high voltage transmission lines. In this case the over current principle cannot easily cope with the change in the direction of the current flow, which is common in the transmission but no so common in radial distribution lines. Computing the impedance in the three-phase system is a bit involved in each type of the fault produces a different impedance expression. Because of these differences the settings of the distance relay are needed to be selected to distinguish between the ground and phase faults.

In addition fault resistance may create problem for distance measurement because of the fault resistance may be difficult for predict. It is particularly challenging for distance relays to measure correct fault impedance when the current in feed from the other end of the line create an unknown voltage drop on the fault resistance (Kazemi, A. et al., 2009), (Kulkami, P.A. et al., 2010).

This may contribute to erroneous computation of the impedance, called apparent impedance 'seen' by the relay located at the end of the line and using the current and voltage measurement just from the end. Once the impedance is computed, it is compared to the settings that define the operating characteristics of the relay. Based on the comparison, a decision is made if a fault has occurred, if so in what zone.

The principle behind the standard distance protection function is based on measured apparent impedance (Z_{seen}) at the transmission line terminals. The apparent impedance is computed from fundamental power frequency components of measured instantaneous voltage and current signals (Liu, Q.; Wang, Z., 2008), (Khederzadeh, M.; Sidhu, T. S., 2006), (Jamali, S.; Shateri, H. 2011), the apparent impedance is given by:

$$Z_{seen} = \left(V_{seen} / I_{seen} \right) . K_Z \tag{16}$$

5. Case study and simulation results

The power system studied in this paper is the 400 kV, 50 Hz eastern Algerian electrical transmission networks at group SONELGAZ (Algerian Company of Electricity and Gas) which is shows in figure 8 (Sonelgaz Group/GRTE, 2011). The MHO distance relay is located in the bus bar at Ramdane Djamel substation in Skikda to protect transmission line between busbar *A* and busbar *B* at Oued El Athmania substation in Mila, the bus bar *C* at Salah Bay substation in Sétif.

The figure below represents a 400 kV transmission line in the presence of a series FACTS type GCSC, TCSC and TCSR installed in the midpoint of the transmission line protected by a MHO distance relay between busbar A and B.

5.1. Characteristic curve of installed series FACTS devices

Figure 9 shows the characteristic curves of the different compensators used GCSC, TCSC and TCSR installed on transmission line in this case study.

5.2. Impact on the impedance of a protected transmission line.

The impact of the angle variation γ and injected reactance X_{GCSC} by compensator GCSC on reactance and resistance of the total impedance for transmission line (X_{AB} and R_{AB}) and on the parameters of measured impedance by MHO distance relay (X_{Relay} and R_{Relay}) in the inductive and capacitive mode is summarized in table 1.

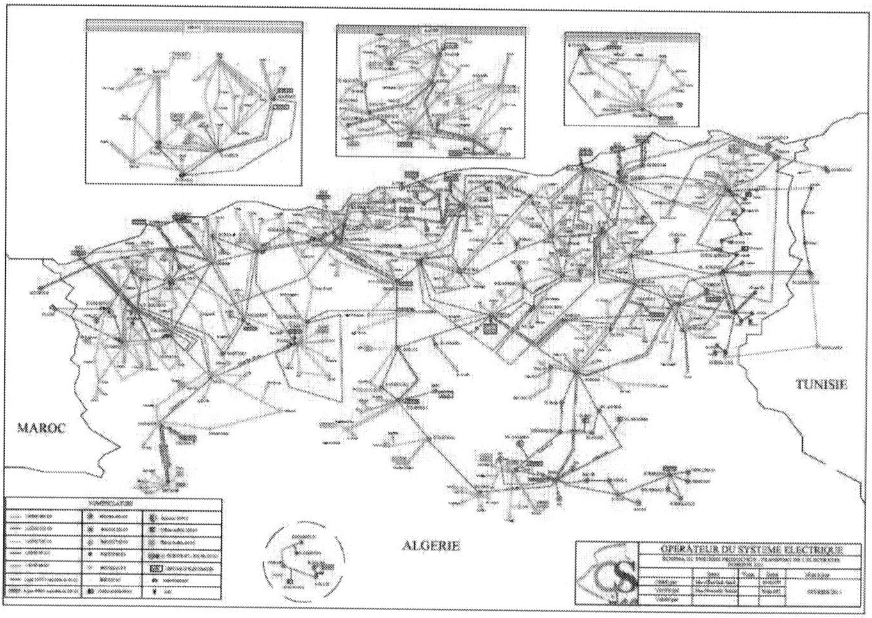

(a). Global.

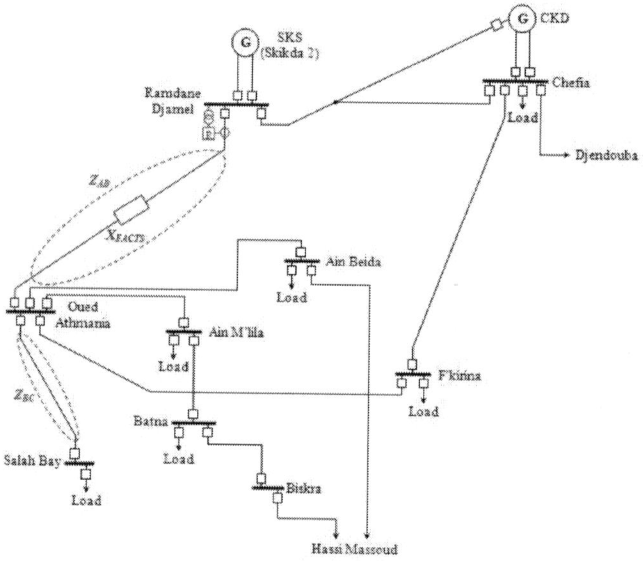

(b). Eastern.

Figure 8. Electrical networks 400 kV study in Algeria

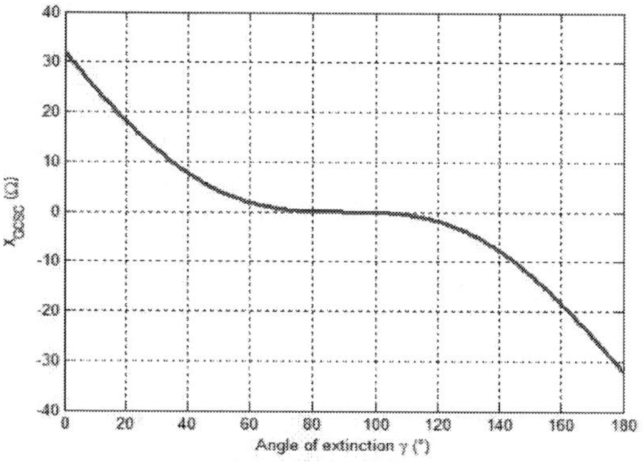

(a). $X_{GCSC} = f(\gamma)$

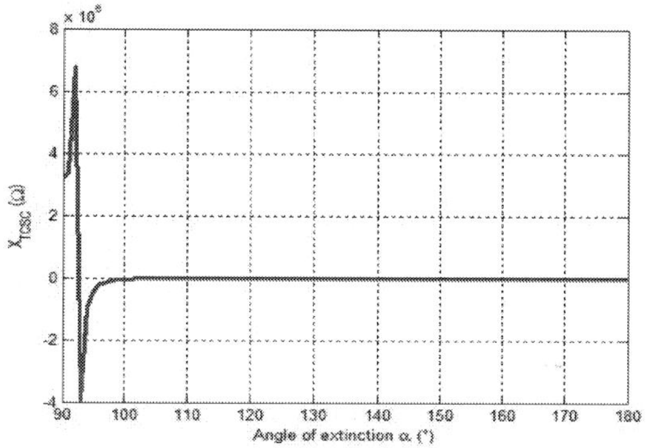

(b). $X_{TCSC} = f(\alpha)$

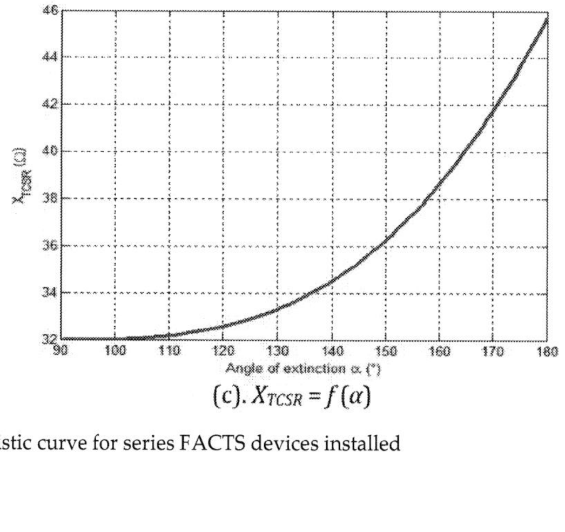

(c). $X_{TCSR} = f(\alpha)$

Figure 9. Characteristic curve for series FACTS devices installed

Mode	Inductive				Capacitive			
γ (°)	0	20	40	80	100	120	140	180
X_{GCSC} (Ω)	32,000	18,3415	7,7466	0,0718	-0,0718	-1,8454	-7,7466	-32,000
X_{AB} (Ω)	143,44	129,78	119,19	111,51	111,37	109,59	103,69	79,440
R_{AB} (Ω)	11,526	11,526	11,526	11,526	11,526	11,526	11,526	11,526
X_{Relay} (Ω)	7,1720	6,4891	5,9593	5,5756	5,5684	5,4797	5,1847	3,9720
R_{Relay} (Ω)	0,5763	0,5763	0,5763	0,5763	0,5763	0,5763	0,5763	0,5763

Table 1. Variation of reactance and resistance as a function of γ and X_{GCSC}

The impact of the angle variation α and X_{TCSC} injected reactance by compensator TCSC on reactance and resistance of the total impedance for transmission line (X_{AB} and R_{AB}) and on the parameters of measured impedance by MHO distance relay (X_{Relay} and R_{Relay}) in the inductive and capacitive mode is summarized in table 2.

Mode	Inductive			Capacitive		
α (°)	90	91	92	100	140	180
X_{TCSC} (Ω)	$3{,}159.10^6$	$3{,}385.10^6$	$6{,}7825.10^6$	-4828,0	-440.684	-106.670
X_{AB} (Ω)	$3{,}158\ 10^6$	$3{,}384.10^6$	$6{,}7826.10^6$	-48177,0	-329,24	4,7697
R_{AB} (Ω)	11,526	11,526	11,526	11,526	11,526	11,526
X_{Relay} (Ω)	$1{,}579.10^5$	$1{,}692.10^5$	$3{,}3913.10^5$	-2408,9	-16.4622	0.2385
R_{Relay} (Ω)	0,5763	0,5763	0,5763	0,5763	0,5763	0,5763

Table 2. Variation of reactance and resistance on function α and X_{TCSC}

The impact of the angle variation α and injected reactance X_{TCSR} by compensator TCSR on reactance and resistance of the total impedance for transmission line (X_{AB} and R_{AB}) and on the parameters of measured impedance by MHO distance relay (X_{Relay} and R_{Relay})in the inductive and capacitive mode is summarized in table 3.

Mode	Inductive							
α (°)	90	100	110	120	130	140	160	180
X_{TCSR} (Ω)	32,000	32,021	32,170	32,563	33,308	34,506	38,645	45,714
X_{AB} (Ω)	143,44	143,46	143,61	144,00	144,75	145,95	150,09	157,15
R_{AB} (Ω)	11,526	11,526	11,526	11,526	11,526	11,526	11,526	11,526
X_{Relay} (Ω)	7,1720	7,1731	7,1805	7.2002	7,2374	7,2973	7,5043	7,8577
R_{Relay} (Ω)	0,5763	0,5763	0,5763	0,5763	0,5763	0,5763	0,5763	0,5763

Table 3. Variation of reactance and resistance on function α and X_{TCSR}

5.3. Impact on setting zones

5.3.1. Impact of GCSC Insertion

Figures 10 and 11 show the impact of the variation extinction angle γ and reactance X_{GCSC} on the value of setting zones reactance and setting zones resistance respectively in presence of GCSC on transmission line.

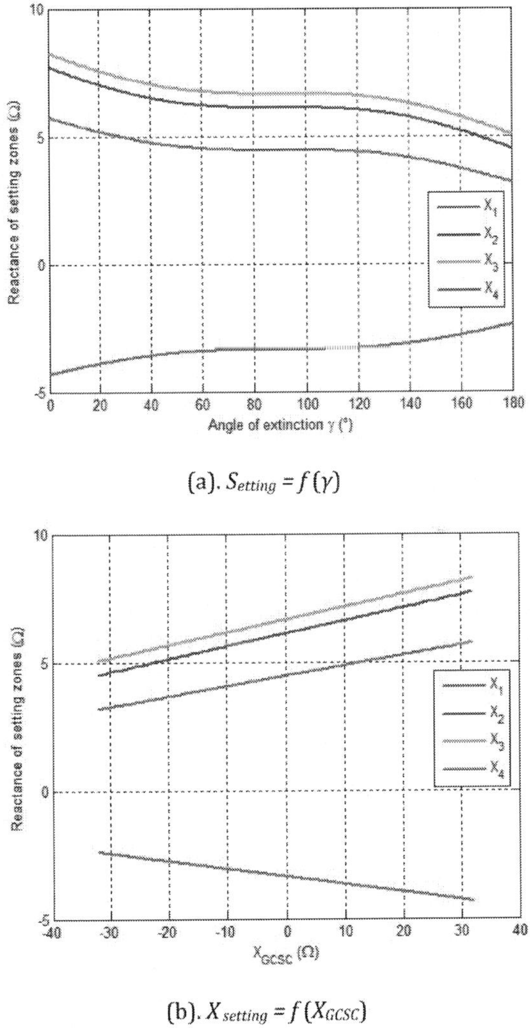

Figure 10. Impact of insertion GCSC on reactance of setting zones

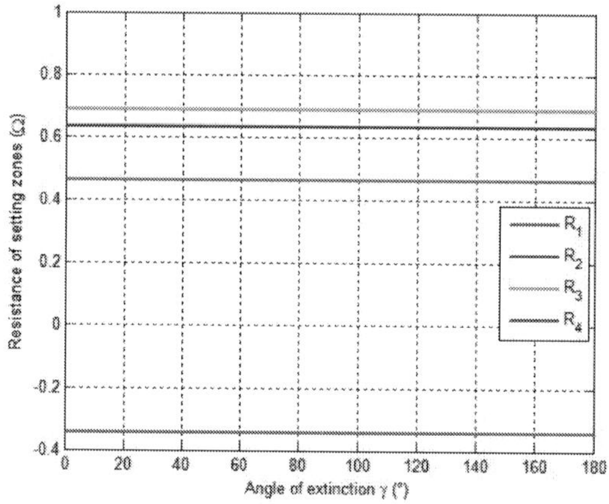

(a). $R_{setting} = f(\gamma)$

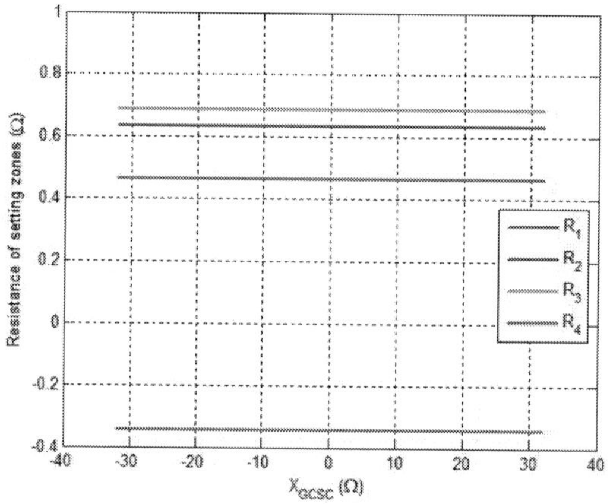

(b). $R_{setting} = f(X_{GCSC})$

Figure 11. Impact of insertion GCSC on resistance of setting zones

5.3.2. Impact of TCSC Insertion

Figures 12 and 13 is show the impact of the variation extinction angle of α and reactance X_{TCSC} on the value of setting zones reactance and setting zones resistance respectively in presence of a TCSC on transmission line.

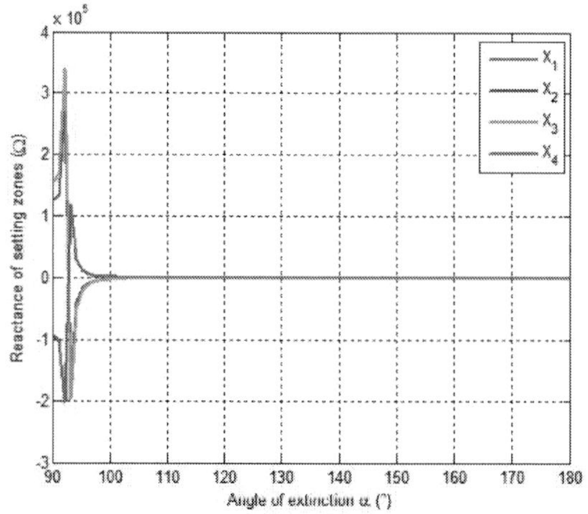

(a). $X_{setting} = f(\alpha)$

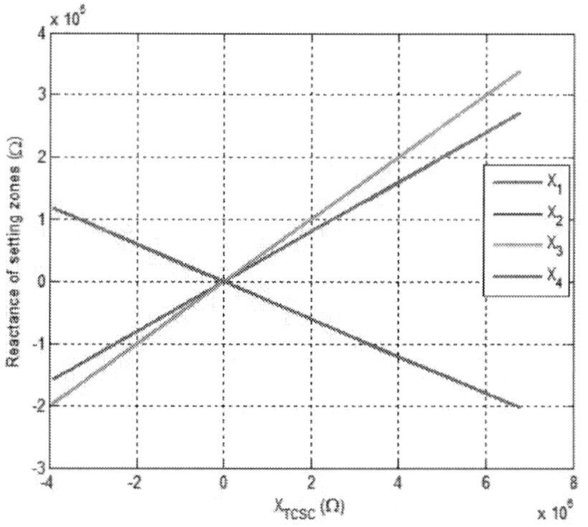

(b). $X_{setting} = f(X_{TCSC})$

Figure 12. Impact of insertion TCSC on reactance of setting zones

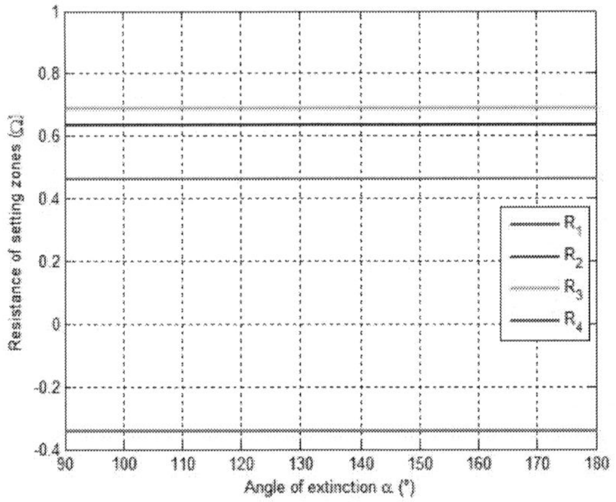

(a). $R_{setting} = f(\alpha)$

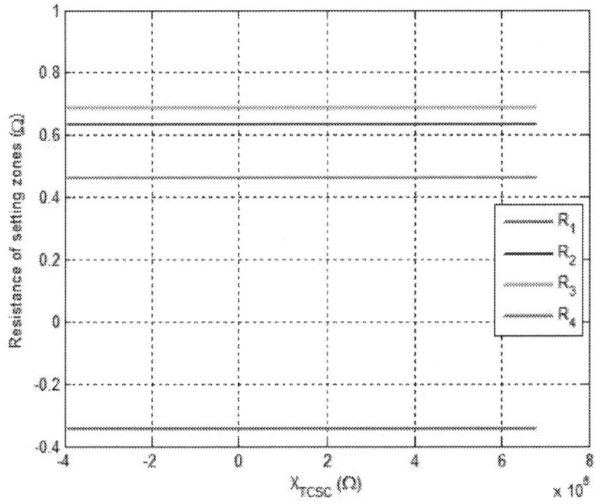

(b). $R_{setting} = f(X_{TCSC})$

Figure 13. Impact of insertion TCSC on resistance of setting zones

5.3.3. Impact of TCSR Insertion

Figures 14 and 15 is show the impact of the variation extinction angle α and reactance X_{TCSR} on the value of setting zones reactance and setting zones resistance respectively in presence of TCSC on transmission line.

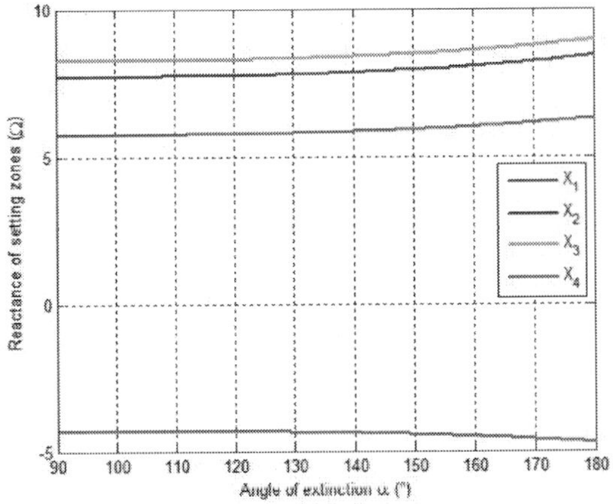

(a). $X_{setting} = f(\alpha)$

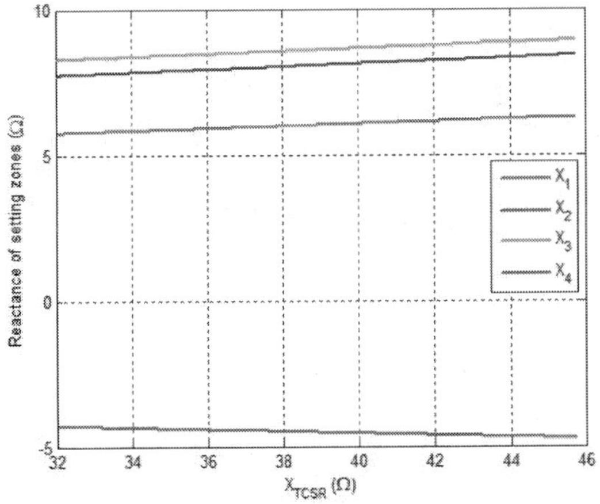

(b). $X_{setting} = f(X_{TCSR})$

Figure 14. Impact of insertion TCSR on reactance of setting zones

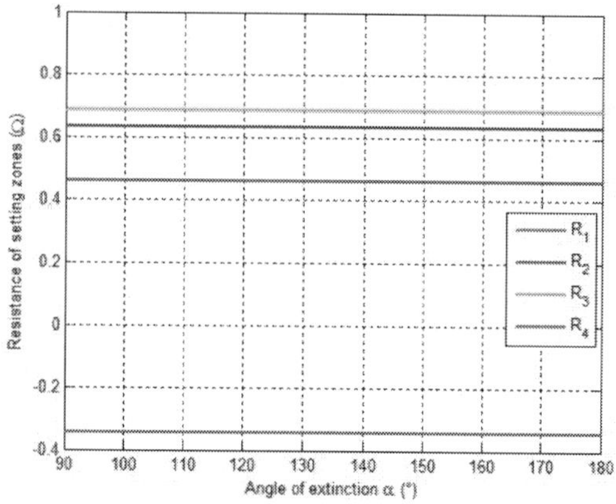

(a). $R_{setting} = f(\alpha)$

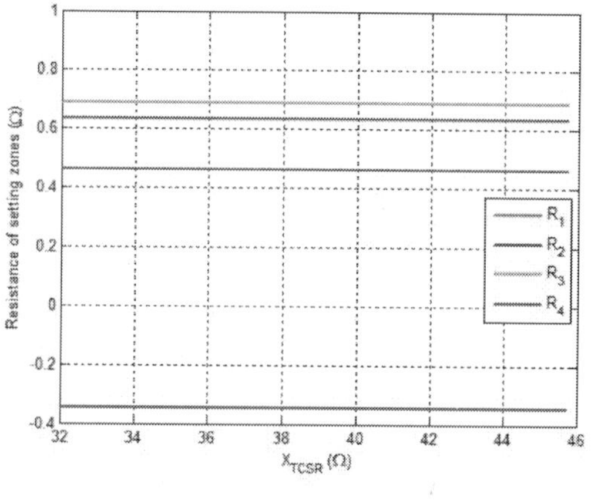

(b) $X_{setting} = f(X_{TCSR})$

Figure 15. Impact of insertion TCSR on resistance of setting zones

6. Conclusions

The results are presented in relation to a typical 400 kV transmission system employing GCSC, TCSC and TCSR series FACTS devices. The effects of the extinction angle γ for controlled GTO installed on GCSC as well as extinction angle α for controlled thyristors on TCSC and TCSR are investigated. These devices are connected at the midpoint of a transmission line protected by distance relay. However as demonstrated these angles injected variable reactance (X_{GCSC}, X_{TCSC} or X_{TCSR}) in the protected line which lead to direct impact on the total impedance of the protected line and setting zones.

Therefore settings zones of the total system protection must be adjusted in order to avoid unwanted circuit breaker tripping in the presence of series FACTS compensator.

Author details

Mohamed Zellagui and Abdelaziz Chaghi
LSP-IE Research Laboratory, Department of Electrical Engineering, Faculty of Technology, University of Batna, Algeria

7. References

Acha, E.; Fuerte-Esquivel, C.R.; Ambriz-Pérez, H.; & Angeles-Camacho, C., (2004). *FACTS Modelling and Simulation in Power Networks*, John Wiley & Sons Ltd Publication, ISBN: 978-0470852712, London, England.

Blackburn, J.L.; Domin, T.J. (2006). *Protective Relaying: Principles and Applications*, 3rd Edition, Published by CRC Press, ISBN: 978-1574447163, USA.

De Jesus F. D.; De Souza L. F. W.; Wantanabe E.; Alves J. E. R. (2007). SSR and Power Oscillation Damping using Gate-Controlled Series Capacitors (GCSC), *IEEE Transaction on Power Delivery*, Vol. 22, N°3, (Mars 2007), pp. 1806-1812.

De Souza, L. F. W.; Wantanabe, E. H.; Alves, J. E. R. (2008). Thyristor and Gate-Controlled Series Capacitors: A Comparison of Component Ratings, *IEEE Transaction on Power Delivery*, Vol. 23, No.2, (May 2008), pp. 899-906.

Dechphung, S.; Saengsuwan, T. (2008). Adaptive Characteristic of MHO Distance Relay for Compensation of the Phase to Phase Fault Resistance, *IEEE International Conference on Sustainable Energy Technologies (ICSET' 2008)*, Singapore, Thailand, 24-27 November 2008.

Gérin-Lajoie, L. (2009), A MHO Distance Relay Device in EMTP Works, *Electric Power Systems Research*, 79(3), March 2009, pp. 484-49.

Horowitz, S.H.; Phadke A.G. (2008). *Power System Relaying*, 3rd Edition, Published by John Wiley & Sons Ltd, ISBN: 978-0470057124, England, UK.

Jamali, S.; Shateri, H. (2011). Impedance based Fault Location Method for Single Phase to Earth Faults in Transmission Systems, *10th IET International Conference on Developments in Power System Protection (DPSP)*, United Kingdom, 29 March - 1 April, 2010.

Kazemi, A.; Jamali, S.; Shateri, H. (2009). Measured Impedance by Distance Relay with Positive Sequence Voltage Memory in Presence of TCSC, *IEEE/PES Power Systems Conference and Exposition (PSCE' 09)*, Seattle, USA, 15-18 March 2009.

Khederzadeh, M.; Sidhu, T. S. (2006). Impact of TCSC on the Protection of Transmission Lines, *IEEE Transactions on Power Delivery*, Vol. 21, No. 1, (January 2006), pp. 80-87.

Kulkami, P. A.; Holmukhe, R. M.; Deshpande, K. D.; Chaudhari, P. S. (2010). Impact of TCSC on Protection of Transmission Line, *International Conference on Energy Optimization and Control (ICEOC' 10)*, Maharashtra, India, 28-30 December 2010.

uses close loop control system to regulate busbar voltage, reactive power exchange, power factor and three phase voltage balance.

This chapter describes a method based on Particle Swarm Optimisation (PSO) [5] to solve the optimal SVC allocation successfully. Particle Swarm Optimisation (PSO) method is a powerful optimization technique analogous to the natural genetic process in biology. Theoretically, this technique is a stochastic approach and it converges to the global optimum solution, provided that certain conditions are satisfied. This chapter considers a distribution system with 9 possible locations for SVCs and 27 different sizes of SVCs. A critical discussion using the example with result is discussed in this chapter.

2. Problem formulation

2.1. Operation principal of SVC

The Static Var Compensator (SVC) are composed of the capacitor banks/filter banks and air-core reactors connected in parallel. The air-core reactors are series connected to thyristors. The current of air-core reactors can be controlled by adjusting the fire angle of thyristors.

The SVC can be considered as a dynamic reactive power source. It can supply capacitive reactive power to the grid or consume the spare inductive reactive power from the grid. Normally, the system can receive the reactive power from a capacitor bank, and the spare part can be consumed by an air-core shunt reactor. As mentioned, the current in the air-core reactor is controlled by a thyristor valve. The valve controls the fundamental current by changing the fire angle, ensuring the voltage can be limited to an acceptable range at the injected node(for power system var compensation), or the sum of reactive power at the injected node is zero which means the power factor is equal to 1 (for load var compensation).

2.2. Assumptions

The optimal SVC placement problem [6] has many variables including the SVC size, SVC cost, locations and voltage constraints on the system. There are switchable SVCs and fixed-type SVCs in practice. However, considering all variables in a nonlinear fashion will make the placement problem very complicated. In order to simplify the analysis, the assumptions are as follows: 1) balanced conditions, 2) negligible line capacitance, 3) time-invariant loads and 4) harmonic generation is solely from the substation voltage supply.

2.3. Radial distribution system

Figure 1 clearly illustrates an m-bus radial distribution system where a general bus i contains a load and a shunt SVC. The harmonic currents introduced by the nonlinear loads are injected at each bus

At the power frequency, the bus voltages are found by solving the following mismatch equations:

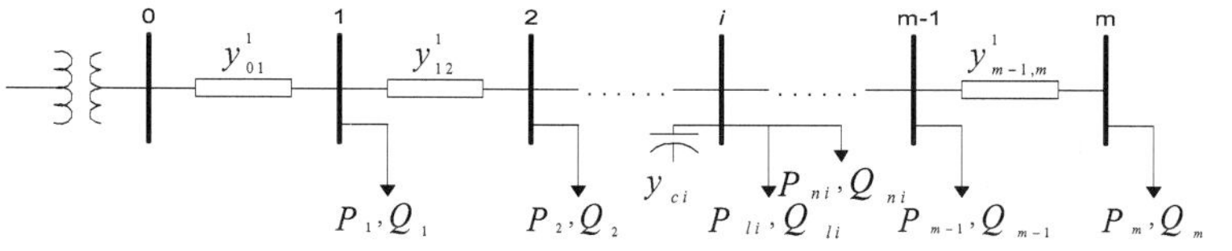

Figure 1. One-line diagram of the radial distribution feeder.

$$P_i = \left|V_i^1\right|^2 G_{ii} + \sum_{\substack{j=1 \\ j\neq i}}^{m} \left|V_i^1 V_j^1 Y_{ij}^1\right| \cos\left(\theta_{ij}^1 + \delta_j^1 - \delta_i^1\right) \quad i = 1,2,3\ldots m \tag{1}$$

$$Q_i = -\left|V_i^1\right|^2 B_{ii} + \sum_{\substack{j=1 \\ j\neq i}}^{m} \left|V_i^1 V_j^1 Y_{ij}^1\right| \sin\left(\theta_{ij}^1 + \delta_j^1 - \delta_i^1\right) \quad i = 1,2,3\ldots m \tag{2}$$

where

$$P_i = P_{li} + P_{ni} \tag{3}$$

$$Q_i = Q_{li} + Q_{ni} \tag{4}$$

$$Y_{ij}^1 = \left|Y_{ij}^1\right| \angle\theta_{ij}^1 = \begin{cases} -y_{ij}^1 & \text{if } i \neq j \\ y_{i-1,i}^1 + y_{i+1,i}^1 + y_{ci}^1 & \text{if } i = j \end{cases} \tag{5}$$

$$Y_{ii} = G_{ii} + jB_{ii} \tag{6}$$

2.4. Real power losses

At fundamental frequency, the real power losses in the transmission line between buses i and i+1 is:

$$P_{loss(i,i+1)}^1 = R_{i,i+1}\left(\left|V_{i+1}^1 - V_i^1\right|\left|Y_{i,i+1}^1\right|\right)^2 \tag{7}$$

So, the total real losses is:

$$P_{loss} = \sum_{n=1}^{N}\left(\sum_{i=0}^{m-1} P_{loss(i,i+1)}^n\right) \tag{8}$$

2.5. Objective function and constraints

The objective function of SVC placement is to reduce the power loss and keep bus voltages and total harmonic distortion (HDF) within prescribed limits with minimum cost. The constraints are voltage limits and maximum harmonic distortion factor, with the harmonics taken into account. Following the above notation, the total annual cost function due to SVC placement and power loss is written as :

Minimize

$$f = K_l K_p P_{loss} + \sum_{j=1}^{m} Q_{cj} K_{cj} \tag{9}$$

where j = 1,2,….m represents the SVC sizes

$$Q_{cj} = j * K_s \tag{10}$$

The objective function (1) is minimized subject to

$$V_{\min} \le |V_i| \le V_{\max} \quad i = 1,2,3\ldots m \tag{11}$$

and

$$HDF_i \le HDF_{\max} \quad i = 1,2,3\ldots m \tag{12}$$

According to IEEE Standard 519 [7] utility distribution buses should provide a voltage harmonic distortion level of less than 5% provided customers on the distribution feeder limit their load harmonic current injections to a prescribed level.

3. Proposed algorithm

3.1. Harmonic power flow [8]

At the higher frequencies, the entire power system is modelled as the combination of harmonic current sources and passive elements. Since the admittance of system components will vary with the harmonic order, the admittance matrix is modified for each harmonic order studied. If the skin effect is ignored, the resulting n-th harmonic frequency load admittance, shunt SVC admittance and feeder admittance are respectively given by:

$$Y_{li}^{n} = \frac{P_{li}}{|V_i^1|^2} - j\frac{Q_{li}}{n|V_i^1|^2} \tag{13}$$

$$Y_{ci}^{n} = nY_{ci}^{1} \tag{14}$$

$$Y_{i,i+1}^{n} = \frac{1}{R_{i,i+1} + jnX_{i,i+1}} \tag{15}$$

The linear loads are composed of a resistance in parallel with a reactance [9]. The nonlinear loads are treated as harmonic current sources, so the injection harmonic current source introduced by the nonlinear load at bus i is derived as follows:

$$I_i^1 = \left[\frac{P_{ni} + jQ_{ni}}{V_i^1} \right]^* \tag{16}$$

$$I_i^n = C(n) I_i^1 \tag{17}$$

In this study, $C(n)$ is obtained by field test and Fourier analysis for all the customers along the distribution feeder. The harmonic voltages are found by solving the load flow equation (18), which is derived from the node equations.

$$Y^n V^n = I^n \tag{18}$$

At any bus i, the r.m.s. value of voltage is defined by

$$|V_i| = \sqrt{\sum_{n=1}^{N} \left| V_i^n \right|^2} \tag{19}$$

where N is an upper limit of the harmonic orders being considered and is required to be within an acceptable range. After solving the load flow for different harmonic orders, the harmonic distortion factor (HDF) [8] that is used to describe harmonic pollution is calculated as follows:

$$HDF_i(\%) = \frac{\sqrt{\sum_{n=2}^{N} \left| V_i^n \right|^2}}{V_i^1} \times 100\% \tag{20}$$

It is also required to be lower than the accepted maximum value.

3.2. Selection of optimal SVC location

The general case of optimal SVC locations can be selected for starting iteration. PSO calculates the optimal SVC sizes according to the optimal SVC locations. After the first time iteration, the solution of SVC locations and sizes will be recorded as old solution and add more locations to consideration. PSO is used to calculate a new solution. If the new solution is better than the old solution, the old solution will be replaced by the new solution. If else, the old solution is the best solution. Therefore, this project will continue to consider more locations until no more optimal solution, which is better than the previous solution. In this chapter, the selection of optimal SVC location is based on the following criteria: voltage, real power loss, load reactive power and harmonic distortion factor with equal weighting.

3.3. Solution algorithm

PSO is a search algorithm based on the mechanism of natural selection and genetics. It consists of a population of bit strings transformed by three genetic operations: 1) Selection or reproduction, 2) Crossover, and 3) mutation. Each string is called chromosome and represents a possible solution. The algorithm starts from an initial population generated randomly. Using the genetic operations considering the fitness of a solution, which corresponds, to the objective function for the problem generates a new generation. The string's fitness is usually the reciprocal of the string's objective function in minimization problem. The fitness of solutions is improved through iterations of generations. For each chromosome population in the given generation, a Newton-Raphson load flow calculation is performed. When the algorithm converges, a group of solutions with better fitness is generated, and the optimal solution is obtained. The scheme of genetic operations, the structure of genetic string, its encode/decode technique and the fitness function are designed. The implementation of PSO components and the neighborhood searching are explained as follows.

4. Implementation of PSO

This section provides a brief introductory concept of PSO. If $X_i = (x_{i1}, x_{i2}, \ldots, x_{id})$ and $V_i = (v_{i1}, v_{i2}, \ldots, v_{id})$ are the position vector and the velocity vector respectively in *d* dimensions search space, then according to a fitness function, where $P_i = (p_{i1}, p_{i2}, \ldots, p_{id})$ is the *pbest* vector and $P_g = (p_{g1}, p_{g2}, \ldots, p_{gd})$ is the *gbest* vector, i.e. the fittest particle of *Pi*, updating new positions and velocities for the next generation can be determined.

```
For each particle
    Initialize particle
END

Do
    For each particle
        Calculate fitness value
        If the fitness value is better than the best fitness
value (pbest), set current value as the new pbest
    End

    Choose the particle with the best fitness value of all
the particles as the gbest
    For each particle
        Calculate particle velocity from equation (21)
        Update particle position from equation (22)
    End

    Perform mutation operation with pm

    While maximum iteration is reached or minimum error
condition is satisfied
```

Figure 2. The pseudo code of the PSO method

$$V_{id}(t) = \omega V_{id}(t-1) + C_1 * rand1 * (P_{id} - X_{id}(t-1)) + C_2 * rand2 * (P_{gd} - X_{id}(t-1)) \quad (21)$$

$$X_{id}(t) = X_{id}(t-1) + V_{id}(t) \quad (22)$$

where V_{id} is the particle velocity, X_{id} is the current particle (solution). P_{id} and P_{gd} are defined as above. *randl* and *rand2* are random numbers which is uniformly distributed between [0,1]. C_1, C_2 are constant values which is usually set to $C_1 = C_2 = 2.0$. These constants represent the weighting of the stochastic acceleration which pulls each particle towards the *pbest* and *gbest* position. ω is the inertia weight and it can be expressed as follows:

$$\omega = (\omega_i - \omega_f) * \frac{iter_{max} - iter}{iter_{max}} + \omega_f \quad (23)$$

whereω_i and ω_f are the initial and final values of the inertia weight respectively. *iter* and $iter_{max}$ are the current iterations number and maximum allowed iterations number respectively.

The velocities of particles on each dimension are limited to a maximum velocity V_{max}. If the sum of accelerations causes the velocity on that dimension to exceed the user-specified V_{max}, the velocity on that dimension is limited to V_{max}.

In this chapter, the parameters used for PSO are as follows:

- Number of particles in the swarm, $N = 30$ (the typical range is 20 – 40)
- Inertia weight, $w_i = 0.9$, $w_f = 0.4$
- Acceleration factor, C_1 and $C_2 = 2.0$
- Maximum allowed generation, $iter_{max} = 100$
- The maximum velocity of particles, V_{max} =10% of search space

There are two stopping criteria in this chapter. Firstly, i(t is the number of iterations since the last change of the best solution is greater than a preset number. The PSO is terminated while maximum iteration is reached.

For the PSO, the constriction and inertia weight factors are introduced and (21) is improved as follows.

$$V_{id}(t) = k\{\omega V_{id}(t-1) + \varphi_1 * rand1 * (P_{id} - X_{id}(t-1)) + \varphi_2 * rand2 * (P_{gd} - X_{id}(t-1))\} \quad (24)$$

$$k = \frac{2}{\left|2 - (\varphi_1 + \varphi_2) - \sqrt{(\varphi_1 + \varphi_2)^2 - 4(\varphi_1 + \varphi_2)}\right|} \quad (25)$$

where k is a constriction factor from the stability analysis which can ensure the convergence (i.e. avoid premature convergence) where $\varphi_1 + \varphi_2 > 4$ and $k_{max} < 1$ and ω is dynamically set as follows:

$$\omega = \omega_{max} - \frac{\omega_{max} - \omega_{min}}{t_{Total}} \times t \qquad (26)$$

where t and t_{Total} is the current iteration and total number of iteration respectively and ω_{min} and ω_{max} is the upper and lower limit which are set 1.3 and 0.1 respectively.

The advantage of the integration of mutation from GAs is to prevent stagnation as the mutation operation choose the particles in the swarm randomly and the particles can move to difference position. The particles will update the velocities and positions after mutation.

$$mutation(x_{id}) = x_{id} - \omega, (-1 < r < 0) \qquad (27)$$

$$mutation(x_{id}) = x_{id} + \omega, \ (1 > r \geq 0) \qquad (28)$$

where x_{id} is a randomly chosen element of the particle from the swarm, ω is randomly generated within the range [0, $\frac{1}{10}$ × (particle $_{max}$ – particle $_{min}$)] (particle $_{max}$ and particle $_{min}$ are the upper and lower boundaries of each particle element respectively) and r is the random number in between 1 and -1

Implementation of an optimization problem is realized within the evolutionary process of a fitness function. The fitness function adopted is derived as equation (9). The objective function is to minimize f. It is composed of two parts; 1) the cost of the power loss in the transmission branch and 2) the cost of reactive power supply. Since PSO is applied to maximization problem, minimization of the problem take the normalized relative fitness value of the population and the fitness function is defined as:

$$f_i = \frac{f_{\max} - f_a}{f_{\max}} \qquad (29)$$

where $f_a = K_l K_p P_{loss} + \sum_{j=1}^{m} Q_{cj} K_{cj}$

5. Software design

Figure 3 depicts the main steps in the process of this chapter. The predefined processes of optimal SVC location and Particle Swarm Optimisation calculation are illustrated in Figure 4 and Figure 5.

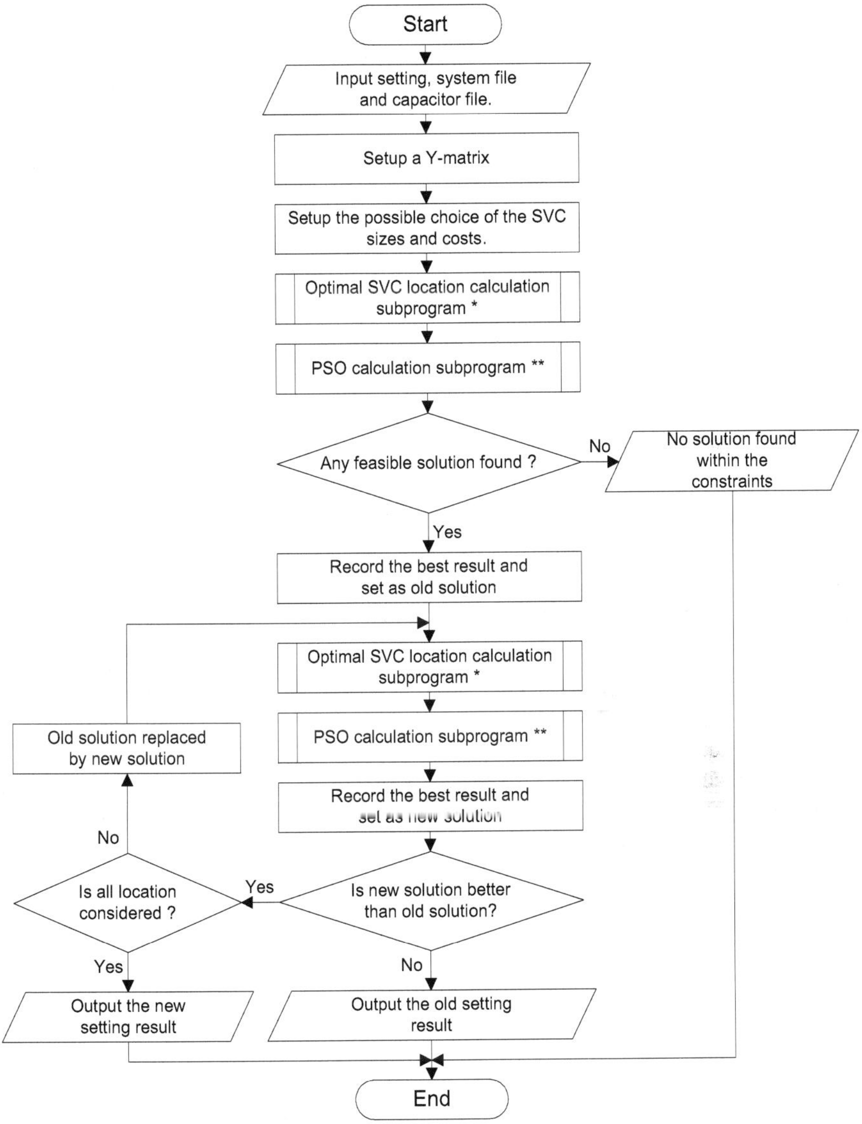

* refer to Figure 4 and ** refer to Figure 5

Figure 3. Flow chart of main operation

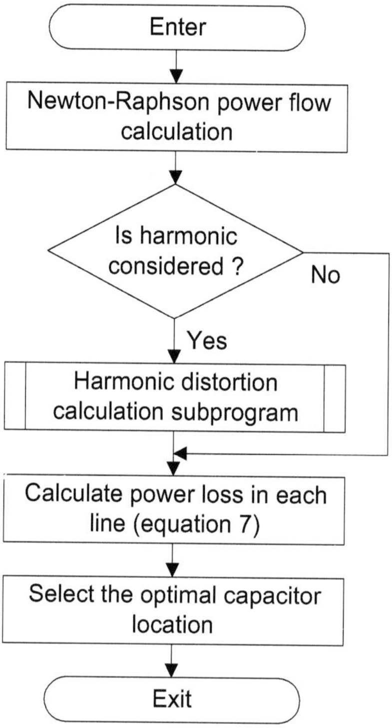

Figure 4. Flow chart of 'Optimal SVC location calculation subprogram' in Figure 2

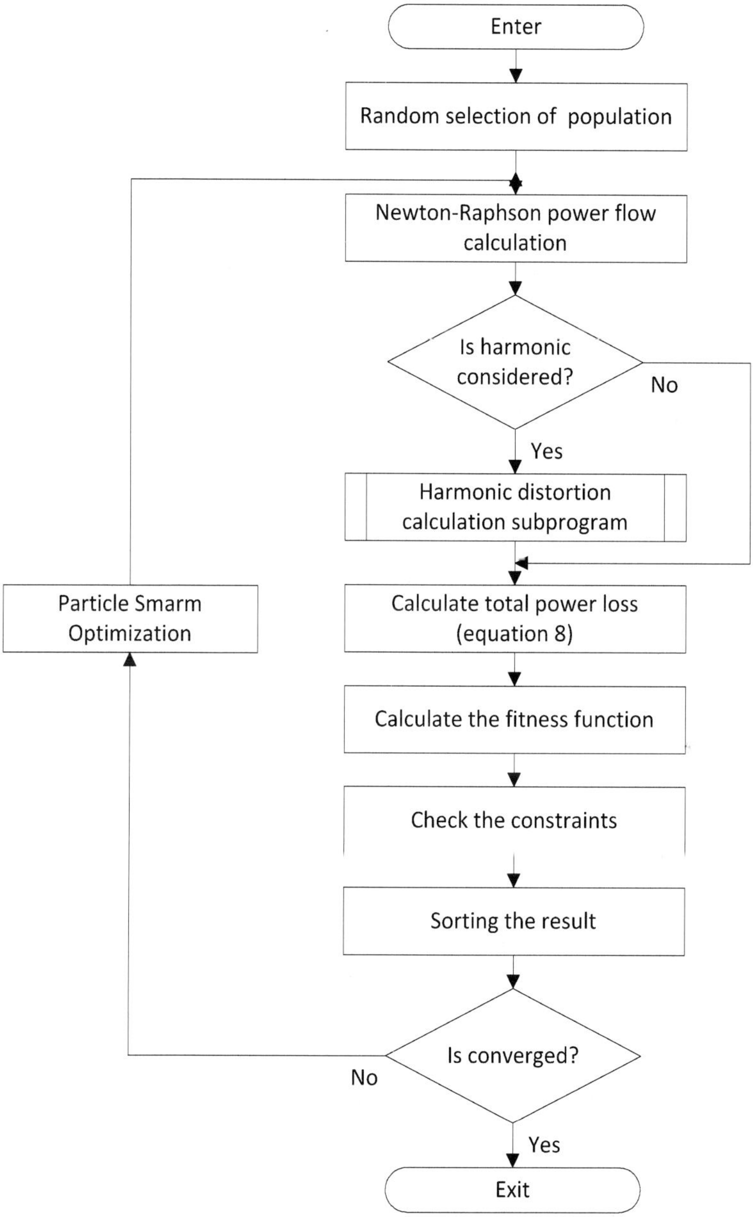

Figure 5. Flow chart of 'Particle Swarm Optimisation calculation subprogram' in Figure 3

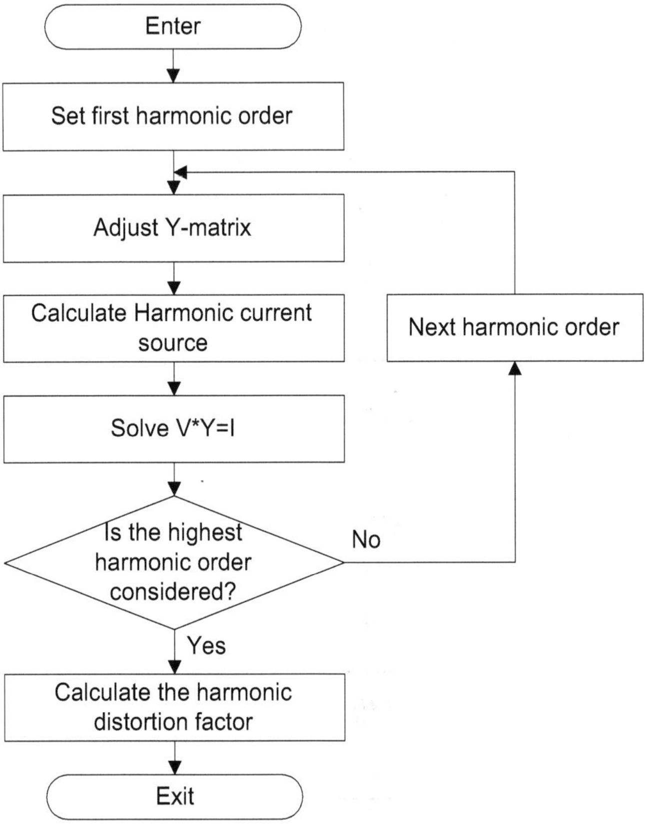

Figure 6. Flow chart of 'Harmonic distortion calculation subprogram' in Figure 4 and Figure 5

6. Numerical example and results

In this section, a radial distribution feeder [10] is used as an example to show the effectiveness of this algorithm. The testing distribution system is shown in Figure 7. This feeder has nine load buses with rated voltage 23kV. Table 1 and Table 2 show the loads and feeder line constants. The harmonic current sources are shown in Table 3, which are generated by each customer.

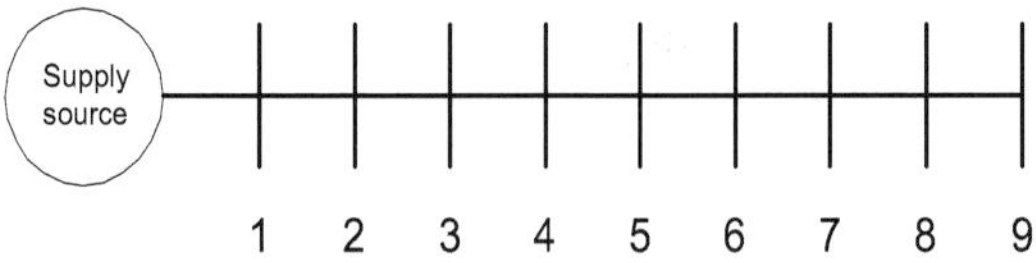

Figure 7. Testing distribution system with 9 buses

Bus	1	2	3	4	5	6	7	8	9
P(kW)	1840	980	1790	1598	1610	780	1150	980	1640
Q(MVAr)	460	340	446	1840	600	110	60	130	200
Non-linear (%)	0	55.7	18.9	92.1	4.7	1.9	38.2	4.5	4.0

Table 1. Load data of the test system

From Bus i	From Bus j	$R_{i,i+1}(\Omega)$	$X_{i,i+1}(\Omega)$
0	1	0.1233	0.4127
1	2	0.0140	0.6051
2	3	0.7463	1.2050
3	4	0.6984	0.6084
4	5	1.9831	1.7276
5	6	0.9053	0.7886
6	7	2.0552	1.1640
7	8	4.7953	2.7160
8	9	5.3434	3.0264

Table 2. Feeder data of the test system

	Harmonic current sources(%) in harmonic order							
Bus	5	7	11	13	17	19	23	25
1	0	0	0	0	0	0	0	0
2	9.1	5.3	1.8	1.1	0.7	0.6	0.4	0.3
3	3.1	1.8	0.6	0.4	0.2	0.2	0.1	0.1
4	6.2	3.6	1.3	0.8	0.5	0.4	0.3	0.2
5	17.7	2.9	4.5	8.2	5.4	2.9	2.9	0
6	0	0	9.6	5.8	0	0	3.6	3.0
7	0.3	0	0	0	0	0	0	0
8	0.8	0.5	0.2	0	0	0	0	0
9	15.1	8.8	3.0	1.8	1.2	1.0	0.6	0.5

Table 3. The harmonic current sources

K_P is selected to be US $168/MW in equation (9). The minimum and maximum voltages are 0.9 p.u. and 1.0 p.u. respectively. All voltage and power quantities are per-unit values. The base value of voltage and power is 23kV and 100MW respectively. Commercially available SVC sizes are analyzed. Table 4 shows an example of such data provided by a supplier for 23kV distribution feeders. For reactive power compensation, the maximum SVC size $Q_{c(max)}$ should not exceed the reactive load, i.e. 4186 MVAr. SVC sizes and costs are shown in Table 5 by assuming a life expectancy of ten years (the placement, maintenance, and running costs are assumed to be grouped as total cost.)

Size of SVC (MVAr)	150	300	450	600	900	1200
Cost of SVC ($)	750	975	1140	1320	1650	2040

Table 4. Available 3-phase SVC sizes and costs

j	1	2	3	4	5	6	7	8	9
Q_{cj} (MVAr)	150	300	450	600	750	900	1050	1200	1350
K_{cj} ($ / MVAr)	0.500	0.350	0.253	0.220	0.276	0.183	0.228	0.170	0.207
j	**10**	**11**	**12**	**13**	**14**	**15**	**16**	**17**	**18**
Q_{cj} (MVAr)	1500	1650	1800	1950	2100	2250	2400	2550	2700
K_{cj} ($ / MVAr)	0.201	0.193	0.187	0.211	0.176	0.197	0.170	0.189	0.187
j	**19**	**20**	**21**	**22**	**23**	**24**	**25**	**26**	**27**
Q_{cj} (MVAr)	2850	3000	3150	3300	3450	3600	3750	3900	4050
K_{cj} ($ / MVAr)	0.183	0.180	0.195	0.174	0.188	0.170	0.183	0.182	0.179

Table 5. Possible choice of SVC sizes and costs

The effectiveness of the method is illustrated by a comparative study of the following three cases. Case 1 is without SVC installation and neglected the harmonic. Both Case 2 and 3 use PSO approach for optimizing the size and the placement of the SVC in the radial distribution system. However, Case 2 does not take harmonic into consideration and Case 3 takes harmonic into consideration. The optimal locations of SVCs are selected at bus 4, bus 5 and bus 9.

Before optimization (Case 1), the voltages of bus 7, 8, 9 are violated. The cost function and the maximum HDF are $132138 and 6.15% respectively. The harmonic distortion level on all buses is higher than 5%.

After optimization (Case 2 and 3), the power losses become 0.007065 p.u. in Case 2 and 0.007036 p.u. in Case 3. Therefore, the power savings will be 0.000747 p.u. in Case 2 and 0.000776 p.u. in Case 3. It can also be seen that Case 3 has more power saving than Case 2.

The voltage profile of Case 2 and 3 are shown in Table 6 and Table 7 respectively. In both cases, all bus voltages are within the limit. The cost savings of Case 2 and Case 3 are $2,744 (2.091%) and $1,904 (1.451%) respectively with respect to Case 1. Since harmonic distortion is considered in Case 3, the sizes of SVCs are larger than Case 2 so that the total cost of Case 3 is higher than Case 2.

The maximum HDF of Case 2 of Case 3 are 1.35% and 1.2% respectively. The HDF improvement of Case 3 with respects to Case 1 is

$$HDF\ improvement\ \% = \frac{6.15 - 1.20}{6.15} \times 100 = 80.49\%$$

The HDF improvement of Case 3 with respects to Case 2 is

$$HDF\ improvement\ \% = \frac{1.40 - 1.20}{1.40} = 14.29\%$$

The improvement of the harmonic distortion is quite attractive and it is clearly shown in Figure 7. The reductions in HDF are 80.49% and 14.29% with respect to Case 1 and Case 2.

The optimal cost and the corresponding SVC sizes, power loss, minimum / maximum voltages, the average CPU time and harmonic distortion factor are also shown in Figure 8.

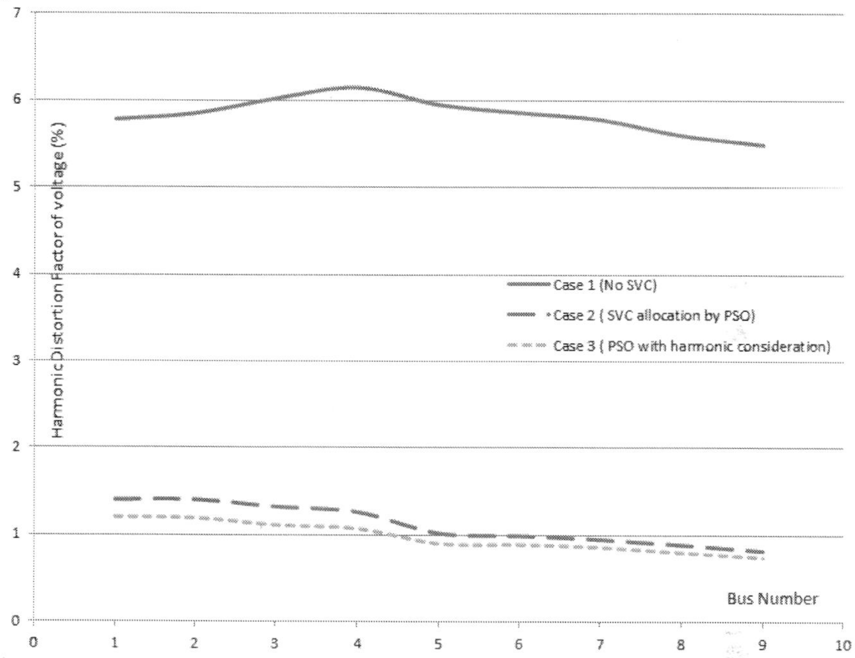

Figure 8. Effect of harmonic distortion on each bus

	Voltages in harmonic order										
	1	5	7	11	13	17	19	23	25	V_{rms}	HDF
Bus	x1	$x10^{-2}$	$x10^{-3}$	$x10^{-3}$	$X10^{-3}$	$x10^{-4}$	$x10^{-4}$	$x10^{-4}$	$x10^{-4}$	x1	%
1	0.993	4.41	2.96	1.57	1.25	9.60	8.12	7.47	4.72	0.992	5.78
2	0.987	4.43	2.98	1.58	1.26	9.69	8.19	7.53	4.76	0.987	5.85
3	0.963	4.45	2.98	1.58	1.26	9.70	8.18	7.54	4.74	0.963	6.02
4	0.948	4.47	3.00	1.59	1.27	9.76	8.21	7.59	4.75	0.947	6.15
5	0.917	4.23	2.78	1.46	1.18	9.02	7.49	6.98	4.24	0.916	5.95
6	0.907	4.14	2.71	1.41	1.14	8.61	7.14	6.65	4.05	0.907	5.86
7	0.889	4.02	2.61	1.34	1.08	8.11	6.72	6.22	3.79	0.888	5.78
8	0.859	3.80	2.43	1.23	0.98	7.31	6.05	5.57	3.40	0.858	5.60
9	0.838	3.66	2.32	1.15	0.91	6.79	5.61	5.13	3.15	0.837	5.49

Table 6. The voltage profile of Case 1

Bus	Voltages in harmonic order									Vrms	HDF
	1	5	7	11	13	17	19	23	25		
	x1	x10^{-2}	x10^{-2}	x10^{-3}	x10^{-3}	x10^{-4}	x10^{-4}	x10^{-4}	x10^{-4}	x1	%
1	0.997	1.190	5.86	1.93	1.22	7.45	6.27	4.49	3.33	0.999	1.40
2	0.999	1.190	5.90	1.94	1.23	7.51	6.32	4.53	3.36	0.988	1.40
3	0.988	1.130	5.34	1.62	0.99	5.51	4.37	2.94	2.05	0.980	1.32
4	0.980	1.100	5.02	1.44	0.85	4.36	3.23	2.05	1.29	0.980	1.26
5	0.962	0.887	3.42	0.81	0.52	2.29	1.33	0.96	0.29	0.962	1.02
6	0.954	0.861	3.28	0.79	0.51	2.12	1.24	1.12	0.49	0.954	0.99
7	0.939	0.827	3.10	0.73	0.46	1.90	1.12	0.97	0.44	0.939	0.95
8	0.915	0.751	2.72	0.60	0.36	1.45	0.89	0.68	0.34	0.915	0.89
9	0.900	0.682	2.37	0.47	0.25	1.04	0.69	0.39	0.25	0.901	0.82

Table 7. The voltage profile of Case 2

Bus	Voltages in harmonic order									Vrms	HDF
	1	5	7	11	13	17	19	23	25		
	x1	x10^{-2}	x10^{-2}	x10^{-3}	x10^{-3}	x10^{-4}	x10^{-4}	x10^{-4}	x10^{-4}	x1	%
1	0.998	1.05	5.08	1.64	1.03	6.41	5.45	3.95	2.98	0.998	1.20
2	1.000	1.06	5.11	1.65	1.04	6.46	5.50	3.98	3.00	1.000	1.19
3	0.991	0.99	4.54	1.33	0.80	4.42	3.53	2.38	1.69	0.991	1.11
4	0.983	0.95	4.20	1.14	0.66	3.25	2.36	1.47	0.91	0.983	1.07
5	0.963	0.81	3.08	0.75	0.52	2.35	1.36	1.05	0.28	0.963	0.90
6	0.955	0.79	2.96	0.74	0.50	2.18	1.26	1.20	0.49	0.955	0.89
7	0.944	0.76	2.81	0.68	0.45	1.95	1.14	1.04	0.44	0.940	0.86
8	0.917	0.69	2.48	0.57	0.35	1.49	0.90	0.73	0.34	0.917	0.80
9	0.902	0.63	2.18	0.45	0.25	1.05	0.69	0.40	0.25	0.902	0.74

Table 8. The voltage profile of Case 3

	Case 1	Case 2	Case 3
Maximum voltage (p.u.)	0.999	0.999	1.000
Minimum voltage (p.u.)	0.837	0.901	0.902
Total power loss (p.u.)	0.007812	0.007065	0.007036
$Q_c(4)$ (p.u.)		0.024	0.036
$Q_c(5)$ (p.u.)		0.024	0.018
$Q_c(9)$ (p.u.)		0.009	0.009
Cost ($ / year)	131238	128494	129334
Average CPU Time (sec.)	0.8	1.20	3.39
Maximum HDF (%)	6.15	1.40	1.20

Table 9. Summary results of the approach

7. Conclusion

This chapter presents a Particle Swarm Optimisation (PSO) approach to searching for optimal shunt SVC location and size with harmonic consideration. The cost or fitness function is constrained by voltage and Harmonic Distortion Factor (HDF). Since PSO is a stochastic approach, performances should be evaluated using statistical value. The performance will be affected by initial condition but PSO can give the optimal solution by increasing the population size. PSO offers robustness by searching for the best solution from a population point of view and avoiding derivatives and using payoff information (objective function). The result shows that PSO method is suitable for discrete value optimization problem such as SVC allocation and the consideration of harmonic distortion limit may be included with an integrated approach in the PSO.

Nomenclature

f_{max}	the maximum fitness of each generation in the population
N	the number of harmonic order is being considered
Q_c	the size of SVC (MVAr)
K_c	the equivalent SVC cost ($/MVAr)
K_l	the duration of the load period
K_p	the equivalent annual cost per unit of power losses ($/kW)
K_s	the SVC bank size (MVAr)
y_{ci}	frequency admittance of the SVC at bus i (pu)
V_i	voltage magnitude at bus i (pu)
P_i, Q_i	active and reactive powers injected into network at bus i (pu)
P_{li}, Q_{li}	linear active and reactive load at bus i (pu)
P_{ni} Q_{ni}	nonlinear active and reactive load at bus i (pu)
θ_{ij}	voltage angle different between bus i and bus j (rad)
G_{ii}, B_{ii}	self conductance and susceptance of bus i (pu)
G_{ij}, B_{ij}	mutual conductance and susceptance between bus i and bus j (pu)

Superscript

1	corresponds to the fundamental frequency value
n	corresponds to the nth harmonic order value

Author details

H.C. Leung and Dylan D.C. Lu
Department of Electrical and Information Engineering,
The University of Sydney, NSW 2006,
Australia

8. References

[1] Ewald Fuchs and Mohammad A. S. Masoum (2008). "Power Quality in Power Systems and Electrical Machines": pp 398-399

[2] Zhang, Wenjuan, Fangxing Li, and Leon M. Tolbert. "Optimal allocation of shunt dynamic Var source SVC and STATCOM: A Survey." 7th IEEE International Conference on Advances in Power System Control, Operation and Management (APSCOM). Hong Kong. 30th Oct.-2nd Nov. 2006.

[3] Verma, M. K., and S. C. Srivastava. "Optimal placement of SVC for static and dynamic voltage security enhancement." International Journal of Emerging Electric Power Systems 2.2 (2005).

[4] Garbex, S., R. Cherkaoui, and A. J. Germond. "Optimal location of multi-type FACTS devices in power system by means of genetic algorithm." IEEE Trans. on Power System 16 (2001): pp 537-544.

[5] Kennedy, J.; Eberhart, R. (1995). "Particle Swarm Optimization". Proceedings of IEEE International Conference on Neural Networks. IV. pp. 1942–1948. http://dx.doi.org/10.1109%2FICNN.1995.488968

[6] Mínguez, Roberto, et al. "Optimal network placement of SVC devices.", IEEE Transactions on Power Systems 22.4 (2007): pp. 1851-1860.

[7] IEEE std. 519-1981, "IEEE Guide for harmonic control and reactive power compensation of static power converters", IEEE, New York, (1981).

[8] J. Arrillaga, D.A. Bradley and P.S. Boodger, "Power system harmonics", John Willey & Sons, (1985), ISBN 0-471-90640-9.

[9] Y. Baghzouz, "Effects of nonlinear loads on optimal capacitor placement in radial feeders", IEEE Trans. Power Delivery, (1991), pp.245-251.

[10] Hamada, Mohamed M., et al. "A New Approach for Capacitor Allocation in Radial Distribution Feeders." The Online Journal on Electronics and Electrical Engineering (OJEEE) Vol. (1) – No. (1), pp 24-29

Chapter 5

Electromechanical Active Filter as a Novel Custom Power Device (CP)

Ahad Mokhtarpour, Heidarali Shayanfar and Mitra Sarhangzadeh

Additional information is available at the end of the chapter

http://dx.doi.org/10.5772/55009

1. Introduction

One of the serious problems in electrical power systems is the increase of electronic devices which are used by the industry as well as residences. These devices, which need high-quality energy to work properly, at the same time, are the most responsible ones for decreasing of power quality by themselves.

In the last decade, Distributed Generation systems (DGs) which use Clean Energy Sources (CESs) such as wind power, photo voltaic, fuel cells, and acid batteries have integrated at distribution networks increasingly. They can affect in stability, voltage regulation and power quality of the network as an electric device connected to the power system.

One of the most efficient systems to solve power quality problems is Unified Power Quality Conditioner (UPQC). It consists of a Parallel-Active Filter (PAF) and a Series-Active Filter (SAF) together with a common dc link [1-3]. This combination allows a simultaneous compensation for source side currents and delivered voltage to the load. In this way, operation of the UPQC isolates the utility from current quality problems of load and at the same time isolates the load from the voltage quality problems of utility. Nowadays, small synchronous generators, as DG source, which are installed near the load can be used for increase reliability and decrease losses.

Scope of this research is integration of UPQC and mentioned synchronous generators for power quality compensation and reliability increase. In this research small synchronous generator, which will be treated as an electromechanical active filter, not only can be used as another power source for load supply but also, can be used for the power quality compensation. Algorithm and mathematical relations for the control of small synchronous generator as an electromechanical active filter have been presented, too. Power quality compensation in sag, swell, unbalance, and harmonized conditions have been done by use of introduced active filter with integration of Unified Power Quality Conditioner (UPQC). In this research, voltage problems are compensated by the Series Active Filter (SAF) of the UPQC. On the other hand, issues related to the compensation of current problems are done by the electromechanical active filter and PAF of UPQC. For validation of the proposed theory in power quality compensation, a simulation has been done in MATLAB/SIMULINK and a number of selected simulation results have been shown.

A T-type active power filter for power factor correction is proposed in [4]. In [5], neutral current in three phase four wire systems is compensated by using a four leg PAF for the UPQC. In [6], UPQC is controlled by H_∞ approach which needs high calculation demand. In [7], UPQC can be controlled based on phase angle control for share load reactive power between SAF and PAF. In [8] minimum active power injection has been used for SAF in a UPQC-Q, based on its voltage magnitude and phase angle ratings in sag conditions. In [9], UPQC control has been done in parallel and islanding modes in dqo frame use of a high pass filter. In [10-12] two new combinations of SAF and PAF for two independent distribution feeders power quality compensation have been proposed. Section 2 generally introduces UPQC. Section 3 explains connection of the proposed active filter. Section 4 introduces electromechanical active filter. Section 5 explains used algorithm for reference generation of the electromechanical filter in detail. Section 6 simulates the paper. Finally, section 7 concludes the results.

2. Unified Power Quality Conditioner (UPQC)

UPQC has composed of two inverters that are connected back to back [2]. One of them is connected to the grid via a parallel transformer and can compensate the current problems (PAF). Another one is connected to the grid via a series transformer and can compensate the voltage problems (SAF). These inverters are controlled for the compensation of the power quality problems instantaneously. Figure 1 shows the general schematic of a UPQC.

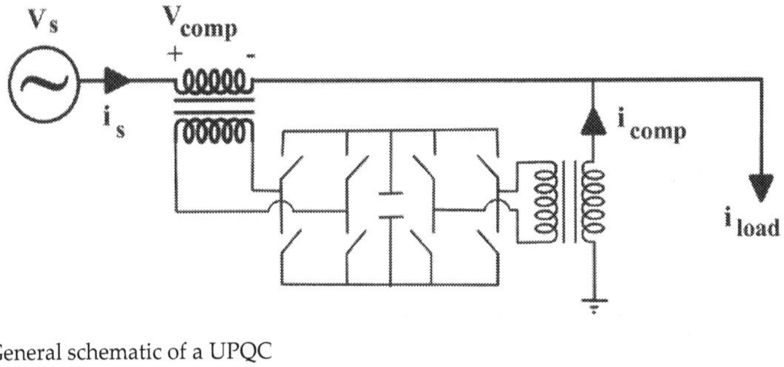

Figure 1. General schematic of a UPQC

A simple circuit model of the UPQC is shown in Figure 2. Series active filter has been modeled as the voltage source and parallel active filter has been modeled as the current source.

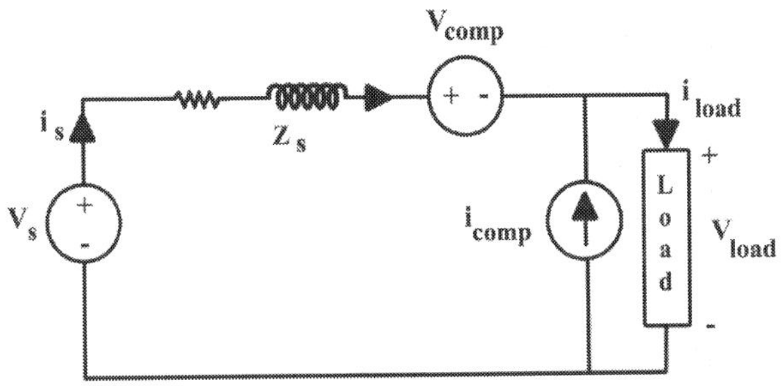

Figure 2. Circuit model of UPQC

3. Connection of Electromechanical Filter

Figure 3, shows schematic of the proposed compensator system. In this research load current harmonics with higher order than 7, has been determined as PAF of UPQC compensator signal. But, load current harmonics with lower order than 7 and reactive power have been compensated by the proposed electromechanical filter.

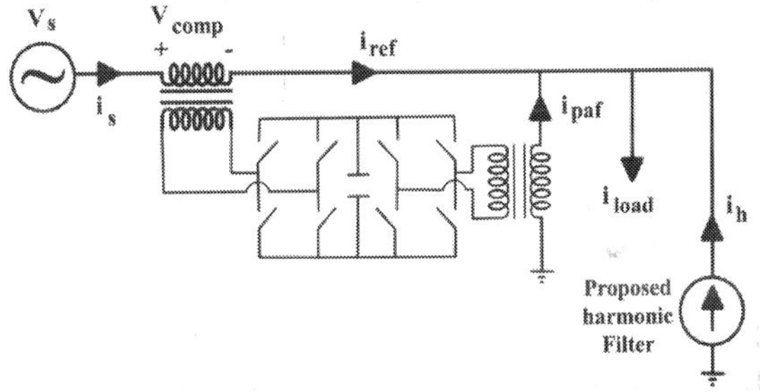

Figure 3. Proposed compensator system

4. Electromechanical Parallel Active Filter

Figure 4, shows the simple structure of a synchronous generator. Based on equation (1), a DC field current of i_f produces a constant magnitude flux.

$$F_f = N_f i_f, \quad \varphi_f = {N_f i_f}/{R}, \quad \psi_f = {N_f N_s i_f}/{R} = M i_f \tag{1}$$

As in [13] N_f and N_s are effective turns of the field windings and the stator windings, respectively; F_f is the magnetomotive force; R is the reluctance of the flux line direction and M is the mutual induction between rotor and stator windings. Speed of rotor is equal to the synchronous speed ($n_s = {120f}/{p}$). Thus, the flux rotates with the angular speed of

$\omega_s = {2\pi n_s}/{60}$. So, stator windings passing flux has been changed as equation (2). It is assumed that in $t = 0$, direct axis of field and stator first phase windings conform each other.

$$\psi_s = i_f M \cos(\omega t) \tag{2}$$

The scope of this section is theoretically investigation of a synchronous machine as a rotating active filter. This theory will be investigated in the static state for a circular rotor type synchronous generator that its equivalent circuit has been shown in Figure 5.

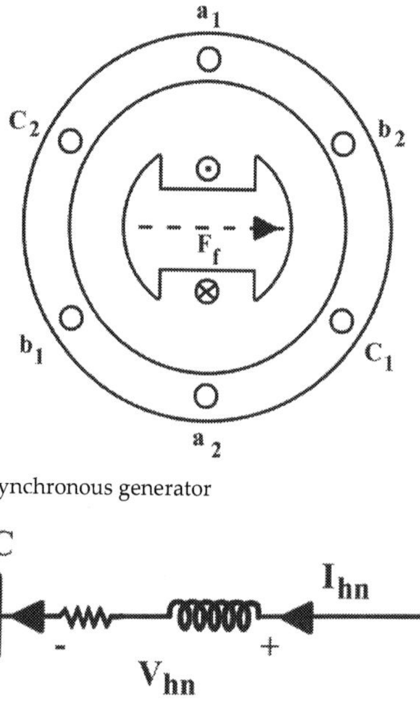

Figure 4. Simple structure of synchronous generator

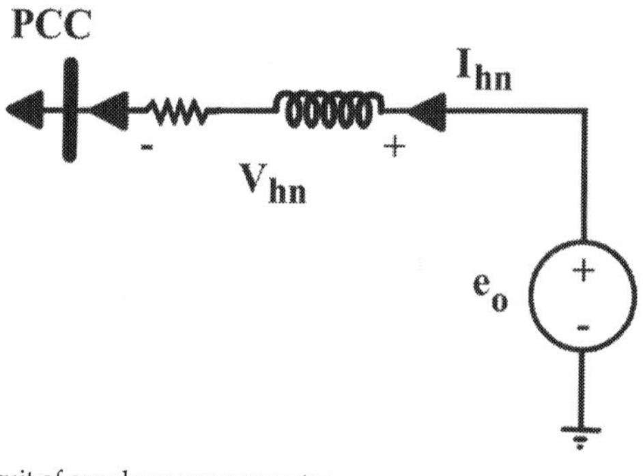

Figure 5. Equivalent circuit of synchronous generator

Equation (3) shows the relation between magnetic flux and voltage behind synchronous reactance of the generator.

$$e = -{d\psi_s(t)}/{dt} = -{d(i_f M \cos(\omega t))}/{dt} = -M {d(i_f \cos(\omega t))}/{dt} \tag{3}$$

Based on equation (3), if the field current be a DC current, the stator induction voltage will be a sinusoidal voltage by the amplitude of $M\omega i_f$. But, if the field current be harmonized as equation (4) then, the flux and internal induction voltage will be as equations (5) and (6), respectively.

$$i_f = I_{dc} + \sum_n I_{fn} \sin(n\omega t - \phi_{fn}) \tag{4}$$

$$\begin{aligned} \psi_f &= i_f M \cos(\omega t) = [I_{dc} + \sum_n I_{fn} \sin(n\omega t - \phi_{fn})] M \cos(\omega t) = \\ & MI_{dc}\cos(\omega t) + \frac{1}{2} M \sum I_{fn} [\sin((n+1)\omega t - \phi_{fn}) + \sin((n-1)\omega t - \phi_{fn})] \end{aligned} \tag{5}$$

$$\begin{aligned} e_o^* &= [-MI_{dc}\omega \sin(\omega t) + \frac{1}{2} MI_{f2}\omega \cos(\omega t - \phi_{f2})] + \\ & \sum_{n=2} [\frac{1}{2} MI_{f(n-1)} n\omega \cos(n\omega t - \phi_{f(n-1)}) + \frac{1}{2} MI_{f(n+1)} n\omega \cos(n\omega t - \phi_{f(n+1)})] \end{aligned} \tag{6}$$

Equation (6) shows that each component of the generator output voltage has composed of two components of the field current. This problem has been shown in Figure 6.

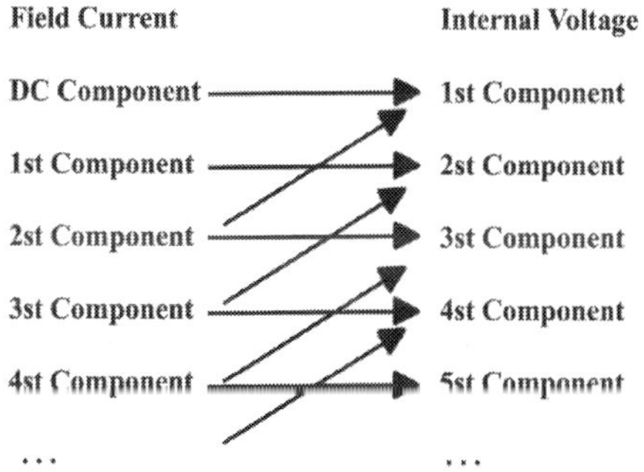

Figure 6. Relation of the field current components by the stator voltage components

It seems that a synchronous generator can be assumed as the Current Controlled System (CCS). Thus it can be used for the current harmonic compensation of a nonlinear load (I_{hn}) as parallel active filter.

5. Algorithm and method

From Figure 5, relation between terminal voltage of the generator and I_{hn} can be derived as equation (7).

$$e_o = V_{pcc} + Z_n I_{hn} = \sum_n V_n \sin(n\omega t + \theta_n) \tag{7}$$

Where, n is the harmonic order; $Z_n = R + jnX$ is the harmonic impedance of the synchronous generator and connector transformer which are known, V_{PCC} is the point of common coupling voltage and I_{hn} is the compensator current that has been extracted from the control circuit.

If similar frequency components of voltage signal e_o^* in equation (6) and e_o in equation (7) set equal, the magnitude and phase angle of the related field current components will be extracted as:

For n=1:

$$-MI_{dc}\omega\sin(\omega t) + \frac{1}{2}MI_{f2}\omega\cos(\omega t - \varphi_{f2}) = V_1\sin(\omega t + \theta_1) \quad (8)$$

$$\sqrt{[-MI_{dc}\omega + \frac{1}{2}MI_{f2}\omega\sin\varphi_{f2}]^2 + [\frac{1}{2}MI_{f2}\omega\cos\varphi_{f2}]^2} = V_1 \quad (9)$$

$$\tan^{-1}[\frac{\frac{1}{2}MI_{f2}\omega\cos\varphi_{f2}}{-MI_{dc}\omega + \frac{1}{2}MI_{f2}\omega\sin\varphi_{f2}}] = \theta_1 \quad (10)$$

For simplicity equations (9) and (10) can be rewritten as follows:

$$X = -MI_{dc}\omega + \frac{1}{2}MI_{f2}\omega\sin\varphi_{f2} \quad (11)$$

$$Y = \frac{1}{2}MI_{f2}\omega\cos\varphi_{f2} \quad (12)$$

$$X^2 + Y^2 = V_1^2 \quad (13)$$

$$X/Y = \theta_1 \quad (14)$$

From the above equations, magnitude and phase of the second component of filed current can result in:

$$X = \frac{V_1}{\sqrt{1 + \tan^2\theta_1}} \quad (15)$$

$$\tan\varphi_{f2} = \frac{X + MI_{dc}\omega}{X\tan\theta_1} \quad (16)$$

$$I_{f2} = \frac{2(X + MI_{dc}\omega)}{M\omega \sin\varphi_{f2}} \tag{17}$$

For n≥2:

$$X = \frac{1}{2}MI_{f(n-1)}n\omega \sin\varphi_{f(n-1)} + \frac{1}{2}MI_{f(n+1)}n\omega \sin\varphi_{f(n+1)} \tag{18}$$

$$Y = \frac{1}{2}MI_{f(n-1)}n\omega \cos\phi_{f(n-1)} + \frac{1}{2}MI_{f(n+1)}n\omega \cos\phi_{f(n+1)} \tag{19}$$

$$X = \frac{V_n}{\sqrt{1 + \tan^2\theta_n}} \tag{20}$$

$$\tan\varphi_{f(n+1)} = \frac{X - 0.5Mn\omega I_{f(n-1)} \sin\varphi_{f(n-1)}}{X\tan\theta_n - 0.5Mn\omega I_{f(n-1)} \cos\varphi_{f(n-1)}} \tag{21}$$

$$I_{f(n+1)} = \frac{2(X - 0.5Mn\omega I_{f(n-1)} \sin\varphi_{f(n-1)})}{Mn\omega \sin\varphi_{f(n-1)}} \tag{22}$$

Where, M and ω are the mutual inductance and angular frequency, respectively. Obviously for the extraction of required components of filed current from the above equations, first suggestion for DC and first order component of the field current are need. Resulted field current can be injected via a PWM and current inverter to the field windings of the synchronous generator. Figure 7, shows the control circuit of the electromechanical

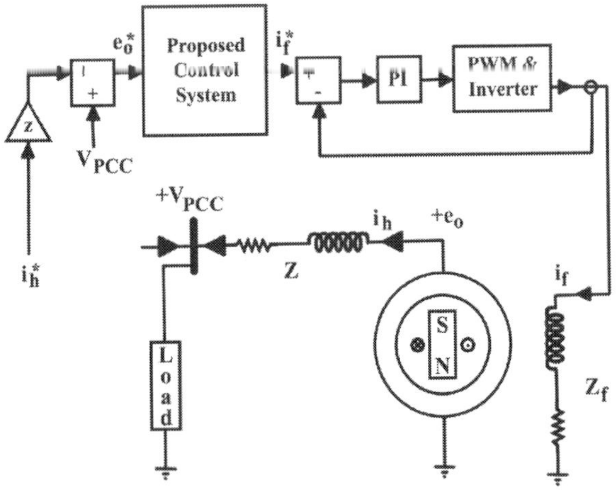

Figure 7. Block diagram of the proposed active filter control

active filter. I_h^* and I_f^* are desired compensator current and calculated field current signal. Detail of the proposed control circuit can be found in the equations (11) to (22). In the present research controlled voltage source of MATLAB has been used instead of required PWM and inverter. Constant and integrator coefficients in the PI controller have been chosen 1000 and 200, respectively. As mentioned earlier first order load active and reactive powers can be easily attended in the electromechanically compensated share of load current for decrease of SAF and PAF power range of UPQC. This problem can control power flow as well as power quality. In other word it can be possible to use a synchronous generator not only for first order voltage generation but, also for the harmonic compensation too.

6. Results

For the investigation of the validity of the mentioned control strategy for power quality compensation of a distribution system, simulation of the test circuit of Figure 8, has been done in MATLAB software. Source current and load voltage, have been measured and analyzed in the proposed control system for the determination of the compensator signals of SAF, PAF and filed current of the electromechanical active filter. Related equations of the controlled system and proposed model of the electromechanical active filter as a current controlled source have been compiled in MATLAB software via M-file. In mentioned control strategy, voltage harmonics have been compensated by SAF of the UPQC and current harmonics with higher order than 7, have been compensated by PAF of UPQC. But, the total of load reactive power, 25 percent of load active power and load current harmonics with lower order than 7 have been compensated by the proposed CCS. This power system consists of a harmonized and unbalanced three phase 380V (RMS, L-L), 50 Hz utility, a three phase balanced R-L load and a three phase rectifier as a nonlinear load. For the investigation of the

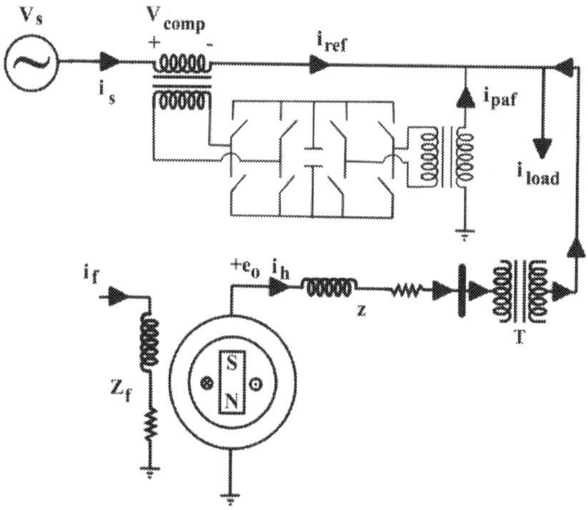

Figure 8. General test system circuit

voltage harmonic condition, utility voltages have harmonic and negative sequence components between 0.05 s and 0.2 s. Also, for the investigation of the proposed control strategy in unbalance condition, magnitude of the first phase voltage is increased to the 1.25 pu between 0.05 s and 0.1 s and decreased to the 0.75 pu between 0.15 s to 0.2 s. Table 1, shows the utility voltage harmonic and sequence parameters data and Table 2, shows the load power and voltage parameters. A number of selected simulation results will be showed further.

Voltage Order	Sequence	Magnitude (pu)	Phase Angle (deg)
5	1	0.12	-45
3	2	0.1	0

Table 1. Utility voltage harmonic and sequence parameters data

Load	Nominal Power (kVA)	Nominal Voltage (RMS, L-L)
Linear	10	380V
Non linear	5	380V

Table 2. Load power and voltage parameters data

Figure 9, shows the source side voltage of phase 1. Figure 10, shows the compensator voltage of phase 1. Figure 11, shows load side voltage of phase 1. Figure 12, shows the load side current of phase 1. Figure 13, shows the CCS current of phase 1 that has been supplied by the proposed active filter. Figure 14, shows the PAF of UPQC current of phase 1. Figure 15, shows the source side current of phase 1. Figure 16, shows the field current of the proposed harmonic filter. Figure 17 and 18 show source voltage and load voltage frequency spectrum, respectively. Figure 19 and 20 show load current and source current frequency spectrum, respectively. Figure 21 and 22 show CCS and PAF frequency spectrum, respectively. Table 3 shows THDs of source and load voltages and currents. Load voltage and source current harmonics have been compensated satisfactory.

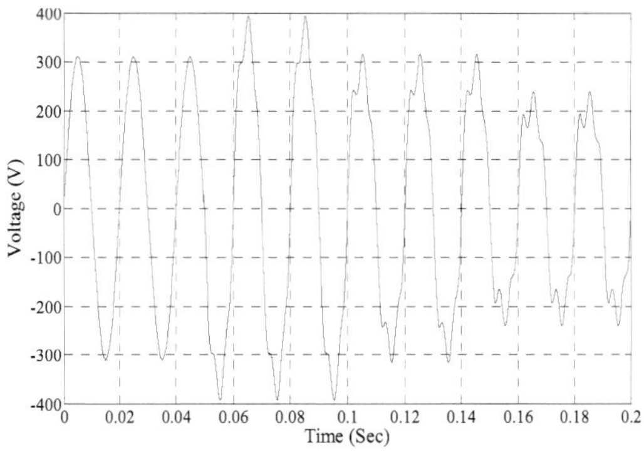

Figure 9. Source side voltage of phase 1 (swell has been occurred between 0.05 and 0.1 sec and sag has been occurred between 0.15 and 0.2 sec. Also, harmonics of positive and negative sequences have been concluded between 0.05 to 0.2 sec)

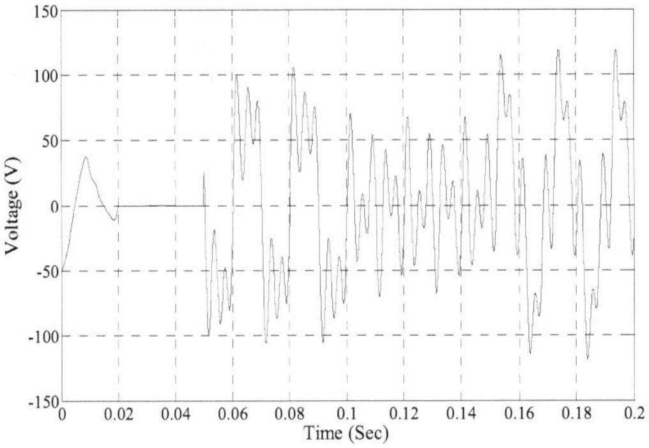

Figure 10. Compensator voltage of phase 1 (compensator voltage has been determined for the sag, swell, negative sequence and harmonics improvement)

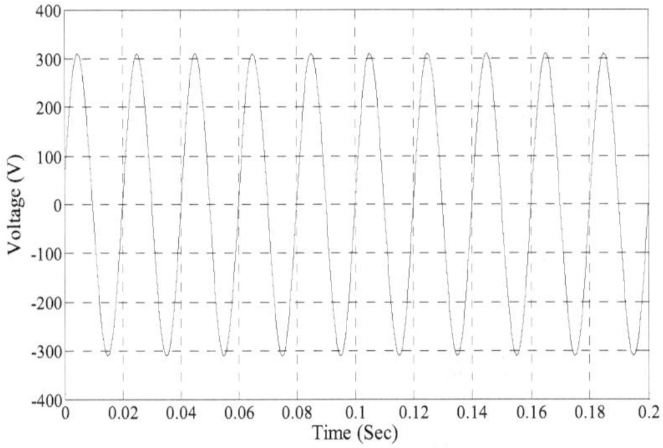

Figure 11. Load side voltage of phase 1 (sag, swell, harmonics, positive and negative sequences have been canceled)

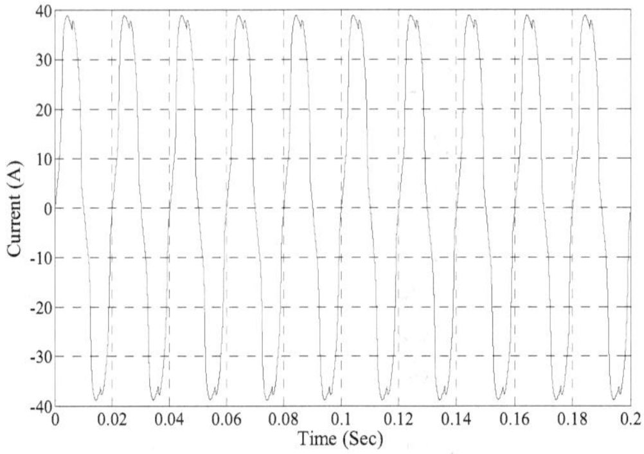

Figure 12. Load side current of phase 1 (it is harmonized. It should be noticed that this current has been calculated after the voltage compensation and thus voltage unbalance has not been transmitted to the current)

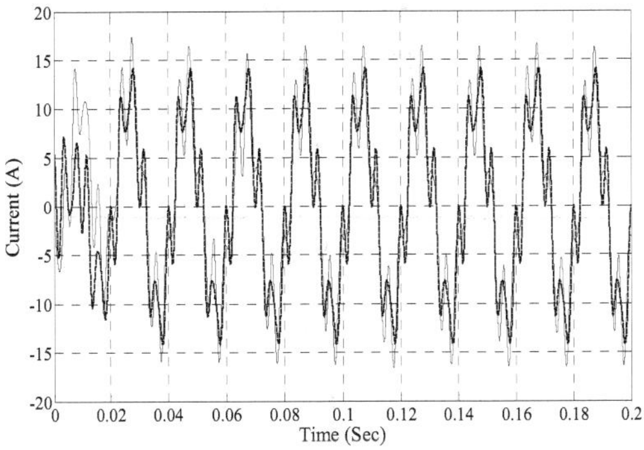

Figure 13. Proposed CCS current of phase 1 (this current has been injected to the grid by the electromechanical active filter. The solid line shows output current of filter and dotted line shows desired current of filter)

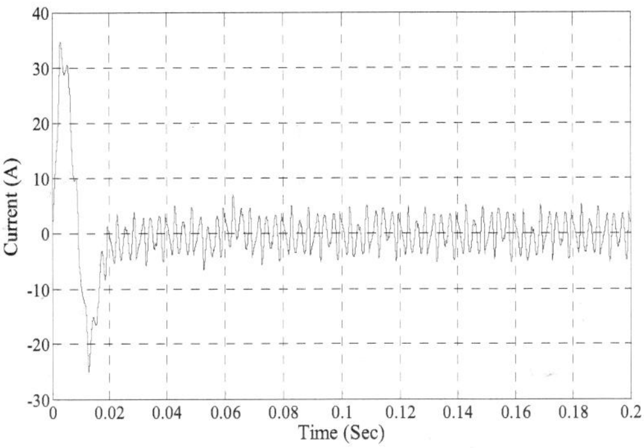

Figure 14. PAF of UPQC current of phase 1 (this current has been injected to the grid by the parallel active filter of UPQC)

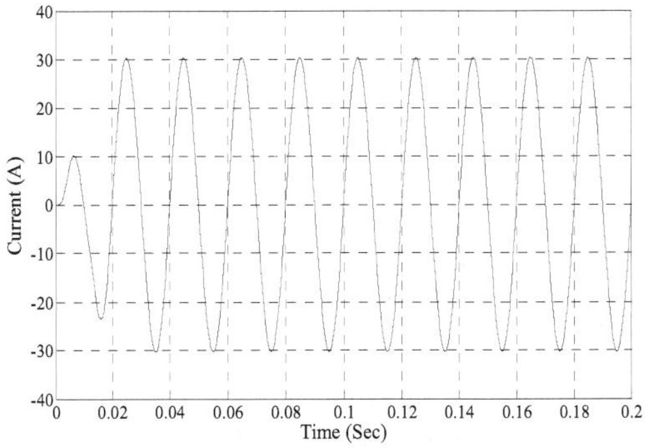

Figure 15. Source side current of phase 1 (harmonics and reactive components of load current have been canceled)

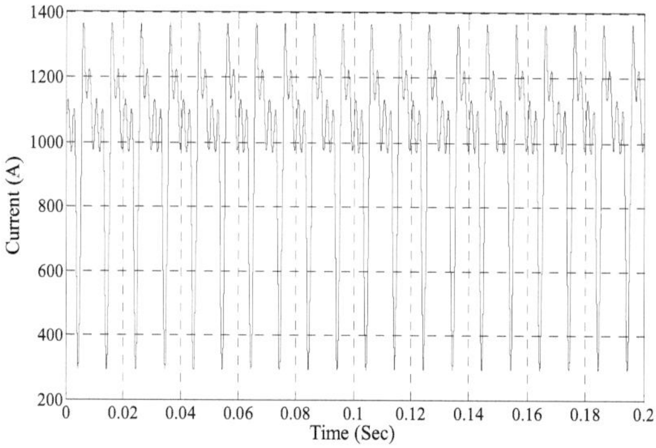

Figure 16. Field current of proposed harmonic filter (field current is controlled for the load active, reactive and harmonic current compensation)

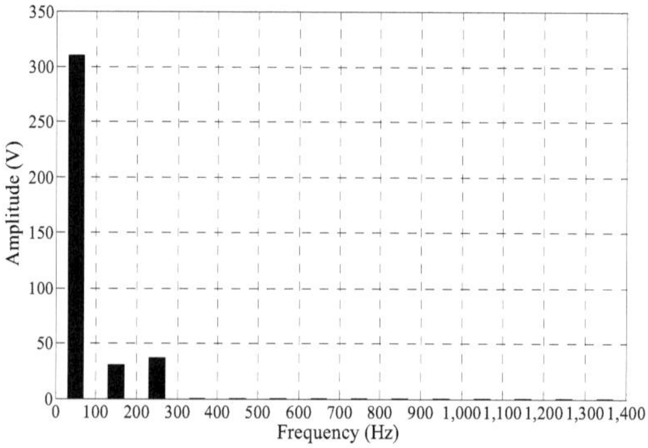

Figure 17. Source side voltage frequency spectrum

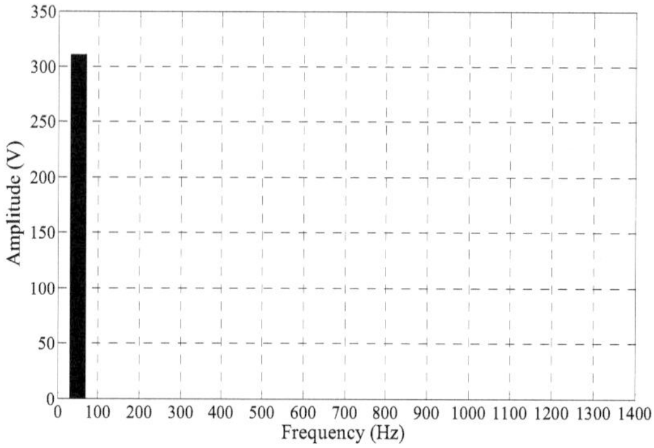

Figure 18. Load side voltage frequency spectrum

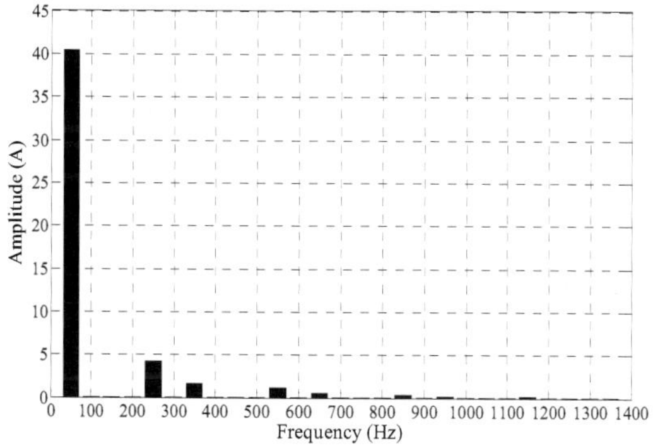

Figure 19. Load side current frequency spectrum

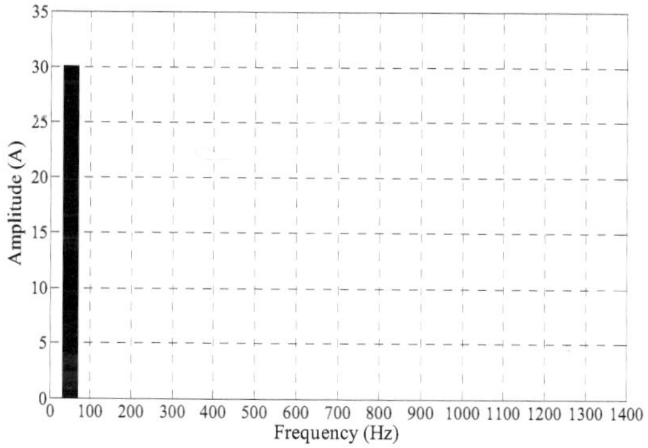

Figure 20. Source side current frequency spectrum

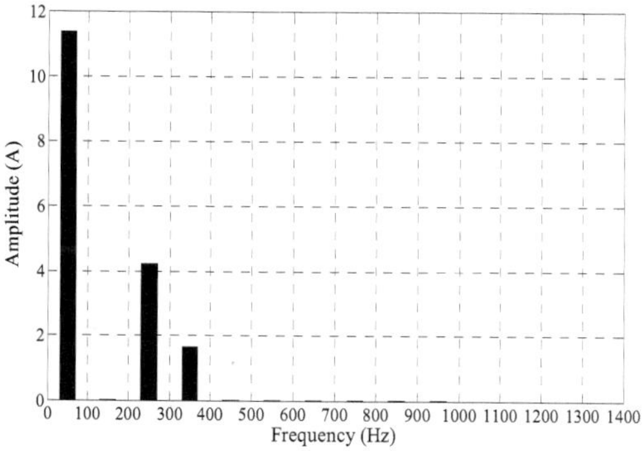

Figure 21. Proposed current controlled system frequency spectrum

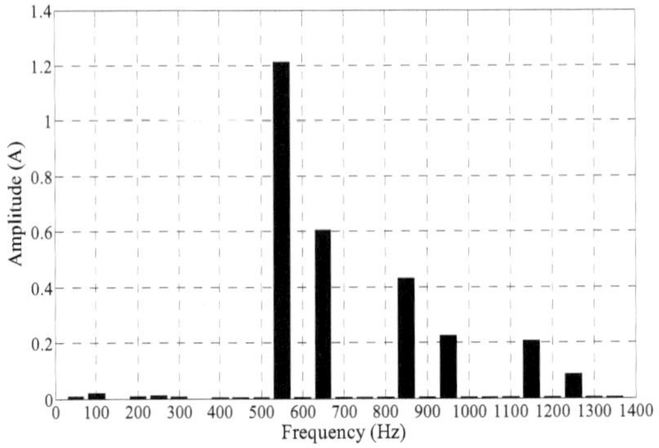

Figure 22. Parallel active filter frequency spectrum

Vs THD	IL THD	VL THD	Is THD
0.1561	0.1179	.001	.0012

Table 3. Total Harmonic Distortion (THD)

7. Conclusions

It is known that use of small synchronous generators in distributed generated networks can reduce transmitted active and reactive powers from the main source and consequently line losses. In this research power quality compensation was done by composition of UPQC and synchronous generators as electromechanical active filter. In other word, by proper determination and control of synchronous generator field current it could be used as controlled current source for power quality compensation. This was for reduction of UPQC power rating in the distributed generated networks. Also, an algorithm was investigated for the determination of the reference field current. Proposed CCS modeling was implemented based on the mentioned related algorithm in MATLAB software. Control strategy had three instantaneously stages. Voltage harmonics were compensated by SAF of the UPQC. Current harmonics with higher order than 7 were compensated by PAF of the UPQC. Lower order current harmonics, load reactive power and a part of load active power were compensated by the proposed controlled current source. Total harmonic distortion of load voltage before compensation was 0.15 which was reduced to almost zero after compensation. Also, total harmonic distortion of the source current before compensation was 0.12 which was reduced to almost zero after compensation.

Author details

Ahad Mokhtarpour, Heidarali Shayanfar and Mitra Sarhangzadeh
Department of Electrical Engineering, Tabriz Branch, Islamic Azad University, Tabriz, Iran

8. References

[1] Fujita H., Akagi H. The Unified Power Quality Conditioner: The Integration of Series and Shunt Active Filters. IEEE Transaction on Power Electronics 1998; 13(2) 315-322.

[2] Shayanfar H. A., Mokhtarpour A. Management, Control and Automation of Power Quality Improvement. In: Eberhard A. (ed.) Power Quality. Austria: InTech; 2010. p127-152.

[3] Hannan M. A., Mohamed A. PSCAD/EMTDC Simulation of Unified Series-Shunt Compensator for Power Quality Improvement. IEEE Transaction on Power Delivery 2005; 20(2) 1650-1656.

[4] Han Y., Khan M.M., Yao G., Zhou L.D., Chen C. A novel harmonic-free power factor corrector based on T-type APF with adaptive linear neural network (ADALINE) control. Simulation Modeling Practice and Theory 2008; 16 (9) 1215–1238.

[5] Khadkikar V., Chandra A. A Novel Structure for Three Phase Four Wire Distribution System Utilizing Unified Power Quality Conditioner (UPQC). IEEE Transactions on Industry Applications 2009; 45(5) 1897-1902.

[6] Kwan K. H., Chu Y.C., So P.L. Model-Based H∞ Control of a Unified Power Quality Conditioner. IEEE Transactions on Industrial Electronics 2009; 56 (7) 2493-2504.

[7] Khadkikar V., Chandra A. A New Control Philosophy for a Unified Power Quality Conditioner (UPQC) to Coordinate Load-Reactive Power Demand between Shunt and Series Inverters. IEEE Transactions on Power Delivery 2008; 23 (4) 2522-2534.

[8] Lee W.C., Lee D.M., Lee T.K. New Control Scheme for a Unified Power Quality Compensator-Q with Minimum Active Power Injection. IEEE Transactions on Power Delivery 2010; 25(2) 1068-1076.

[9] Han B., Bae B., Kim H., Baek S. Combined Operation of Unified Power-Quality Conditioner with Distributed Generation. IEEE Transaction on Power Delivery 2006; 21(1) 330-338.

[10] Mohammadi H.R., Varjani A.Y., Mokhtari H. Multiconverter Unified Power-Quality Conditioning System: MC-UPQC. IEEE Transactions on Power Delivery 2009; 24(3) 1679-1686.

[11] Jindal A.K., Ghosh A., Joshi A. Interline Unified Power Quality Conditioner. IEEE Transactions on Power Delivery 2007; 22(1) 364-372.

[12] Mokhtarpour A., Shayanfar H.A., Tabatabaei N.M. Power Quality Compensation in two Independent Distribution Feeders. International Journal for Knowledge, Science and Technology 2009; 1 (1) 98-105.

[13] Machowski J., Bialek J., Bumby J.R. Power System Dynamics and Stability, United Kingdom: John Wiley and Sons; 1997.

Chapter 6

Reference Generation of Custom Power Devices (CPs)

Ahad Mokhtarpour, Heidarali Shayanfar and Seiied Mohammad Taghi Bathaee

Additional information is available at the end of the chapter

http://dx.doi.org/10.5772/54680

1. Introduction

One of the serious problems in electrical power systems is the increase of electronic devices which are used by the industry. These devices, which need high-quality energy to work properly, at the same time, are the most responsible ones for decreasing of power quality by themselves.

Custom power devices (CP) used in distribution systems can control power quality. One of the most efficient CPs is Unified Power Quality Conditioner (UPQC). It consists of a Parallel-Active Filter (PAF) and a Series-Active Filter (SAF) together with a common dc link [1-3]. This combination allows a simultaneous compensation for source side currents and delivered voltage to the load. In this way, operation of the UPQC isolates the utility from current quality problems of the load and at the same time isolates the load from the voltage quality problems of utility.

Reference generation of UPQC is an important problem. One of the scopes of this research is extending of Fourier transform for increasing of its responsibility speed twelve times as the main control part of reference generation of the UPQC. Proposed approach named Very Fast Fourier Transform (VFFT) can be used in balanced three phase systems for extraction of reference voltage and current signals. Proposed approach has fast responsibility as well as good steady state response. As it is known, Fourier transform response needs at least one cycle data for the settling down which results in slow responsibility and week capability in dynamic condition. In the proposed approach there are two different data window lengths. In the sag or swell condition, control system switches to T/12 data window length but, in the steady state condition it is switched to T/2 data window length. It causes fast responsibility as well as good steady state response. This approach will be used for the UPQC control circuit for extraction of the reference signals.

Second scope of this research is to use Multy Output ADAptive LINEar (MOADALINE) approach for the reference generation of UPQC. Simplicity and flexibility in extraction of different reference signals can be advantage of the proposed algorithm. Third scope of this research is reference generation of UPQC with the scope of power flow control as well as power quality compensation. In this stage, SAF is controlled by dqo approach for voltage sag, swell, unbalance, interruption, harmonic compensation and power flow control. Also, PAF is controlled by composition of dqo and Fourier theories for current harmonic and reactive power compensation.

Also for the validity of the proposed approaches, power quality compensation has been done in a test circuit via simulation. Voltage sag, swell and harmonics will be compensated by SAF of UPQC. Also, current harmonics and reactive power will be compensated by PAF of UPQC. Section 2 generally introduces UPQC and its equivalent circuit. Section 3 explains the proposed VFFT and related equations. Section 4 introduces UPQC reference generation system based on the proposed VFFT approach. Section 5 explains proposed MOADALINE algorithm for reference generation. Section 6 explains reference generation based on power flow control. Section 7 simulates the research. Finally, section 8 concludes the results.

2. Unified Power Quality Conditioner (UPQC)

UPQC is composed of two inverters that are connected back to back [3-10]. One of them is connected to the grid via a parallel transformer and can compensate the current problems (PAF). Another one is connected to the grid via a series transformer and can compensate the voltage problems (SAF). These inverters are controlled for the compensation of the power quality problems instantaneously. Figure 1 shows the general schematic of a UPQC.

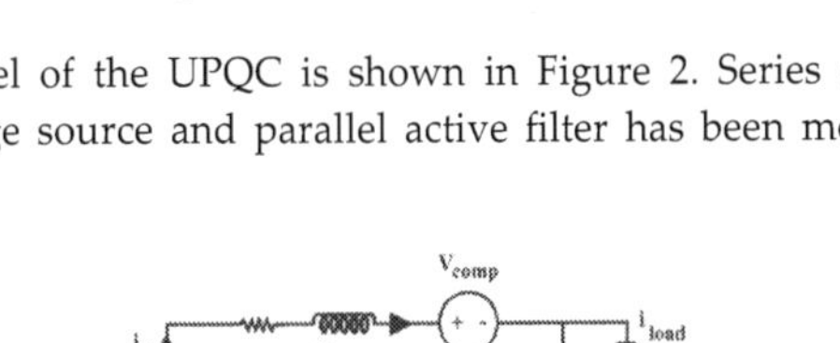

Figure 1. General schematic of a UPQC

A simple circuit model of the UPQC is shown in Figure 2. Series active filter has been modeled as the voltage source and parallel active filter has been modeled as the current source.

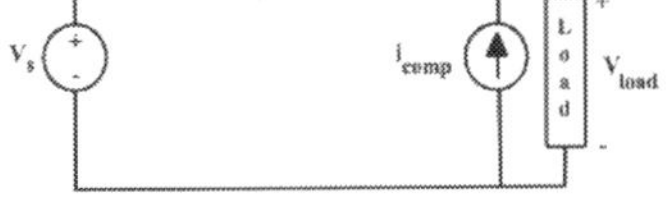

Figure 2. Circuit model of UPQC

3. VFFT problem statement

Fourier transform has the capability of different order components extraction of distorted periodic voltage and current. It is possible to use voltage and current first order components for determining the compensator signals. Based on the related equations of Fourier transform, there is a need for at least one cycle data for settling down the response. This problem can cause week responsibility in dynamic condition. Proposed extended Fourier transform will be responsible for improving this problem. First order Fourier coefficients of a sinusoidal signal can be written as equations (1) and (2).

$$a_1 = \frac{2}{2\pi}\int_0^{2\pi} v(\omega t)\cos(\omega t)d\omega t = \frac{4}{2\pi}\int_0^{\pi} v(\omega t)\cos(\omega t)d\omega t \tag{1}$$

$$b_1 = \frac{2}{2\pi}\int_0^{2\pi} v(\omega t)\sin(\omega t)d\omega t = \frac{4}{2\pi}\int_0^{\pi} v(\omega t)\sin(\omega t)d\omega t \tag{2}$$

Based on equations (1) and (2) it is possible to reduce settling time of the Fourier transform responsibility to T/2; where, T is the main period of the signal. In this condition the responsibility speed will be increased twice but, it is not reasonable speed in dynamic condition yet. Equation (5) can be resulted from equations (3) and (4), for a sinusoidal signal with phase angle Φ. Equation (6) can be resulted similarly.

$$\frac{4}{2\pi}\int_0^{\pi} \sin(\omega t+\phi)\cos(\omega t)d\omega t = \sin\phi \tag{3}$$

$$\frac{8}{2\pi}\int_0^{\frac{\pi}{6}} \sin(\omega t+\phi)\cos(\omega t)d\omega t + \frac{8}{2\pi}\int_{\frac{2\pi}{6}}^{\frac{3\pi}{6}} \sin(\omega t+\phi)\cos(\omega t)d\omega t + \frac{8}{2\pi}\int_{\frac{4\pi}{6}}^{\frac{5\pi}{6}} \sin(\omega t+\phi)\cos(\omega t)d\omega t = \sin\phi \tag{4}$$

$$\frac{4}{2\pi}\int_0^{\pi} \sin(\omega t+\phi)\cos(\omega t)d\omega t = \frac{8}{2\pi}\int_0^{\frac{\pi}{6}} \sin(\omega t+\phi)\cos(\omega t)d\omega t + \frac{8}{2\pi}\int_{\frac{2\pi}{6}}^{\frac{3\pi}{6}} \sin(\omega t+\phi)\cos(\omega t)d\omega t + \frac{8}{2\pi}\int_{\frac{4\pi}{6}}^{\frac{5\pi}{6}} \sin(\omega t+\phi)\cos(\omega t)d\omega t = \sin\phi \tag{5}$$

$$\frac{4}{2\pi}\int_0^{\pi} \sin(\omega t+\phi)\sin(\omega t)d\omega t = \frac{8}{2\pi}\int_0^{\frac{\pi}{6}} \sin(\omega t+\phi)\sin(\omega t)d\omega t + \frac{8}{2\pi}\int_{\frac{2\pi}{6}}^{\frac{3\pi}{6}} \sin(\omega t+\phi)\sin(\omega t)d\omega t + \frac{8}{2\pi}\int_{\frac{4\pi}{6}}^{\frac{5\pi}{6}} \sin(\omega t+\phi)\sin(\omega t)d\omega t = \cos\phi \tag{6}$$

Equations (7) and (8) can be rewritten from equations (5) and (6) respectively.

$$a_1 = \frac{8}{2\pi}\int_0^{\frac{\pi}{6}} \sin(\omega t+\phi)\cos(\omega t)d\omega t + \frac{8}{2\pi}\int_0^{\frac{\pi}{6}} (-\sin(\omega t+\phi-{2\pi}/{3}))(-\cos(\omega t-{2\pi}/{3}))d\omega t +$$
$$\frac{8}{2\pi}\int_0^{\frac{\pi}{6}} \sin(\omega t+\phi+{2\pi}/{3})\cos(\omega t+{2\pi}/{3})d\omega t = \frac{8}{2\pi}\int_0^{\frac{\pi}{6}} \sin(\omega t+\phi)\cos(\omega t)d\omega t + \tag{7}$$
$$\frac{8}{2\pi}\int_0^{\frac{\pi}{6}} \sin(\omega t+\phi+{2\pi}/{3})\cos(\omega t+{2\pi}/{3})d\omega t + \frac{8}{2\pi}\int_0^{\frac{\pi}{6}} \sin(\omega t+\phi-{2\pi}/{3})\cos(\omega t-{2\pi}/{3})d\omega t$$

$$b_1 = \frac{8}{2\pi}\int_0^{\frac{\pi}{6}} \sin(\omega t+\phi)\sin(\omega t)d\omega t + \frac{8}{2\pi}\int_0^{\frac{\pi}{6}} (-\sin(\omega t+\phi-{2\pi}/{3}))(-\sin(\omega t-{2\pi}/{3}))d\omega t +$$
$$\frac{8}{2\pi}\int_0^{\frac{\pi}{6}} \sin(\omega t+\phi+{2\pi}/{3})\sin(\omega t+{2\pi}/{3})d\omega t = \frac{8}{2\pi}\int_0^{\frac{\pi}{6}} \sin(\omega t+\phi)\sin(\omega t)d\omega t + \tag{8}$$
$$\frac{8}{2\pi}\int_0^{\frac{\pi}{6}} \sin(\omega t+\phi+{2\pi}/{3})\sin(\omega t+{2\pi}/{3})d\omega t + \frac{8}{2\pi}\int_0^{\frac{\pi}{6}} \sin(\omega t+\phi-{2\pi}/{3})\sin(\omega t-{2\pi}/{3})d\omega t$$

This solution has advantage of reducing the data window length to T/12. In this condition, settling time of the Fourier transform responsibility will be decreased to T/12. In other words, the responsibility speed will be increased twelve times. It should be noticed that this approach can be used only in balanced three phase systems. In this condition reach to the same accuracy in response is possible as compared to that of one cycle but very faster than one cycle. When data window length is chosen T sec, all of even components of distorted voltage and current in the related integral of Fourier transform can be filtered completely. But this is not possible in T/12 data window length condition. So, the problem of this approach is the unfiltered steady state oscillations in the response. This is because of small data window length. Thus, to complete the proposed approach for accessing the fast response, as well as good steady state response, proper composition of T/12 and T/2 data window lengths have been used. Completed proposed approach uses T/12 data window length in signal magnitude change instances, for accessing the fast response, and T/2 data window length in other times, for accessing the good response without oscillation [10].

4. UPQC reference generation based on the VFFT

Based on the previous indications, the detection of sag, swell, and load change conditions are essential problem for proper control of the UPQC. In this research, for the detection of sag or swell condition in the source voltage and changes in the load current, derivative of the first order component magnitude of the voltage and current signals has been compared with a constant value. This is based on the facts that in these conditions, the derivative of the signal magnitude can change rapidly with an almost increasing slope amount. Figure 3, shows the block diagram of the proposed VFFT based control approach [10].

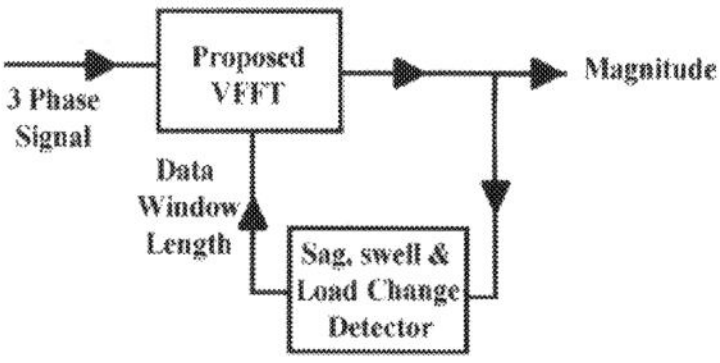

Figure 3. Proposed VFFT based control system

4.1. Series active filter reference generation

Proposed approach can be used for the extraction of direct and indirect first order components of the source voltage. In this control approach, series active filter can compensate harmonics as well as voltage sag and swells. In this approach the reference voltage magnitude can be setup to the nominal value as equation (9). Figure 4, shows the block diagram of the SAF control circuit.

$$v_1(t) = a_1\cos(\omega t) + b_1\sin(\omega t) = \sqrt{a_1^2 + b_1^2}\sin(\omega t + \arctan\frac{a_1}{b_1})$$
$$v_{ref}(t) = v_{nom}\sin(\omega t + \arctan\frac{a_1}{b_1}) = v_{nom}\sin(\omega t + \phi_{vl1}) \tag{9}$$

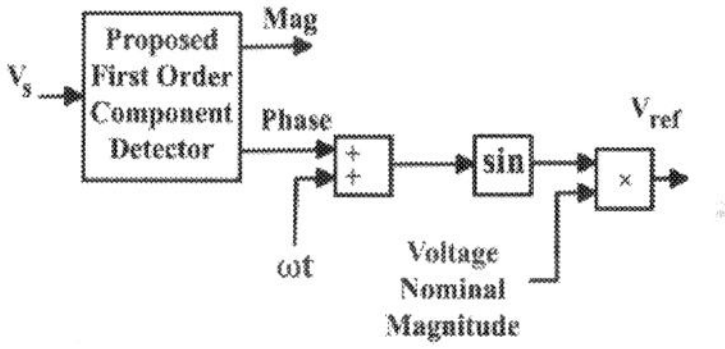

Figure 4. SAF control system block diagram

4.2. Parallel active filter reference generation

In this research parallel active filter have the duty of the reactive power compensation, as well as, current harmonics. For this purpose as equation (10), active first order component of load current which is tangent component of the load current to the load voltage can be used as the reference current. This problem has been shown in Figure 5.

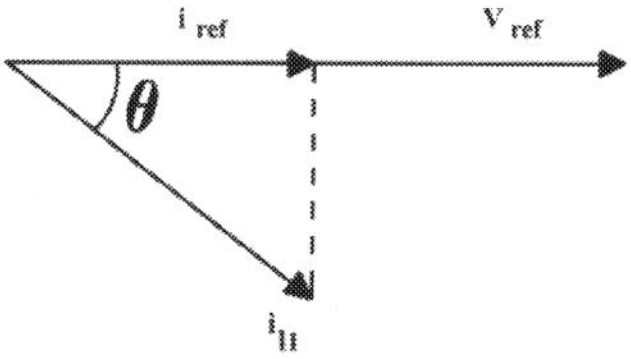

Figure 5. Reference current determination

$$I_{ref}(t) = I_{l1}\cos(\phi_{vl1} - \phi_{il1})\sin(\omega t + \phi_{vl1}) \tag{10}$$

Where, I_{l1} and Φ_{il1} are the first order component magnitude and phase angle of the load current, respectively and Φ_{vl1} is the first order component phase angle of the load voltage. Figure 6, shows the block diagram of the PAF control circuit.

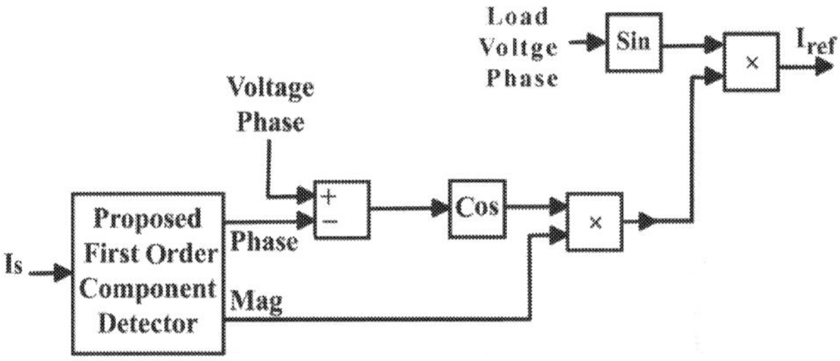

Figure 6. Block diagram of the PAF control circuit

5. Reference generation based on MOADALINE approach

Based on MOADLINE approach each $n \times 1$ signal of y can be written as a weighted linear combination of its components. If S(t) be $n \times m$ component matrix of y at time t and W(t) be $m \times 1$ vector of weighted coefficient then, the signal of y can be written as equation (11). MOADALINE can be used for determining weight vector of W which generates a special signal of y from its components [11]. Weighted factors can be updated in each stage of an adaptation approach for extraction of a desired signal. Equation (11) shows adaptation rule that is based on Least Mean Square (LMS) algorithm.

$$\begin{aligned} y &= SW \\ W(t+dt) &= W(t) + kS^T(t)[S(t)S^T(t)]^{-1}e(t) \end{aligned} \tag{11}$$

Where, e(t) is the error between desired and actual signal of y, and K is the convergence factor. It is possible to extract the reference voltage and current from uncompensated source voltage and load current. Fourier coefficients can be determined as vector of y. Reference signal can be determined as W. Matrix of S is constant. After determination of uncompensated signal Fourier coefficients, they can be compared with the desired values. Error signal can be used in adaptation rule for updating the vector of W. Therefore reference voltage and current can be determined. Figure 7, shows block diagram of the proposed MOADALINE approach [11].

$$y = \begin{bmatrix} a_n \\ b_n \\ a_0 \end{bmatrix} \tag{12}$$

$$S(t) = \frac{2}{m-1}\begin{bmatrix} \cos(n\omega t_0) & \cos(n\omega t_1) & \dots & \cos(n\omega t_{m-1}) \\ \sin(n\omega t_0) & \sin(n\omega t_1) & \dots & \sin(n\omega t_{m-1}) \\ \frac{1}{2} & \frac{1}{2} & \dots & \frac{1}{2} \end{bmatrix} \quad (13)$$

$$W = \begin{bmatrix} w(t_0) \\ w(t_1) \\ \dots \\ w(t_{m-1}) \end{bmatrix} \quad (14)$$

$$\begin{bmatrix} a_n \\ b_n \\ a_0 \end{bmatrix} = \frac{2}{m-1}\begin{bmatrix} \cos(n\omega t_0) & \dots & \cos(n\omega t_{m-1}) \\ \sin(n\omega t_0) & \dots & \sin(n\omega t_{m-1}) \\ \frac{1}{2} & \dots & \frac{1}{2} \end{bmatrix}\begin{bmatrix} w(t_0) \\ w(t_1) \\ \dots \\ w(t_{m-1}) \end{bmatrix} \quad (15)$$

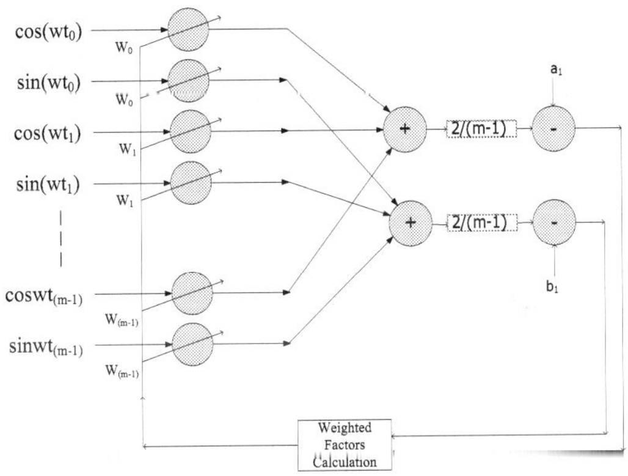

Figure 7. Block diagram of the proposed MOADALINE approach

6. Reference generation based on power flow control

Park transform which is used for the conversion of abc to dqo frame, is useful in the steady state and dynamic analysis of electrical systems [3]. Conversion matrix and related equations have been shown in Equations (16) and (17).

$$\begin{bmatrix} v_d \\ v_q \\ v_o \end{bmatrix} = \frac{1}{3}\begin{bmatrix} 2\sin(\omega t) & 2\sin(\omega t - \frac{2\pi}{3}) & 2\sin(\omega t + \frac{2\pi}{3}) \\ 2\cos(\omega t) & 2\cos(\omega t - \frac{2\pi}{3}) & 2\cos(\omega t + \frac{2\pi}{3}) \\ 1 & 1 & 1 \end{bmatrix}\begin{bmatrix} v_a \\ v_b \\ v_c \end{bmatrix} = \overline{T}\begin{bmatrix} v_a \\ v_b \\ v_c \end{bmatrix} \quad (16)$$

$$\begin{bmatrix} v_a \\ v_b \\ v_c \end{bmatrix} = \overline{T^{-1}} \begin{bmatrix} v_d \\ v_q \\ v_o \end{bmatrix} \quad (17)$$

Instantaneous active and reactive powers in dqo axis can be written as Equations (18) and (19).

$$P = \frac{3}{2}(v_d i_d + v_q i_q + 2v_0 i_0) \quad (18)$$

$$Q = \frac{3}{2}(v_q i_d - v_d i_q) \quad (19)$$

Equations (20) and (21) show direct and quadratic axis voltages based on P and Q that have been determined from equations (18) and (19). It should be noticed that the active and reactive powers in equations (20) and (21) are transmitted powers after series active filter toward the load. These equations show that in constant impedance loads there is a relation between voltage and transmitted power. In other words, a particular load voltage is needed for a particular load power and vice versa. In equations (20) and (21) it have been assumed that the voltages are balanced and $v_0 = 0$.

$$v_d = \frac{2}{3}\left(\frac{Pi_d - Qi_q}{i_d^2 + i_q^2}\right) \quad (20)$$

$$v_q = \frac{2}{3}\left(\frac{Pi_q + Qi_d}{i_d^2 + i_q^2}\right) \quad (21)$$

Therefore, setup transmission active and reactive powers in equations (20) and (21) in considered amounts will result the related load voltage magnitude and phase angle. In reactive power compensated condition, Q=0 and the amount of active power will be extracted from the above equations as equation (22).

$$P = \frac{3}{2}\sqrt{(v_d^2 + v_q^2)(i_d^2 + i_q^2)} \quad (22)$$

Generally, in this approach nominal active and reactive powers of load will be substituted in equation (20) and (21) for reference voltage extraction. But, there is a problem. If we don't have load nominal power data, extracted reference voltage will not have nominal magnitude. For correct arrangement of the voltage magnitude, a PI controller is used for minimum error between magnitudes of extracted reference voltage and the nominal voltage.

$$P_{error} = k_p(V_{setup} - V_{ref}) + k_i \int (V_{setup} - V_{ref})dt \quad (23)$$

Where, V_{ref} and V_{setup} are the nominal voltage magnitude and extracted voltage magnitude, respectively.

Figure 8, shows the block diagram of the proposed control circuit.

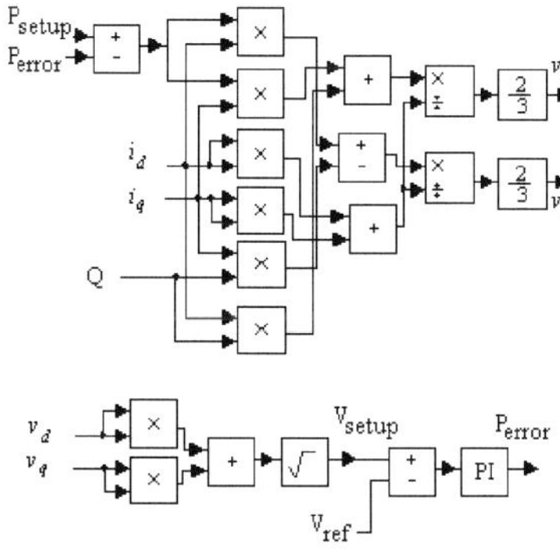

Figure 8. Block diagram of the voltage control circuit

7. Results

7.1. Results of VFFT approach

For the investigation of the validity of the proposed VFFT reference generation strategy in power quality compensation of a distribution system, simulation of the test circuit of Figure 9, has been done in MATLAB/SIMULINK software. Source voltage and load current, have been measured and analyzed in the proposed control system for the determination of the compensator signals of SAF and PAF.

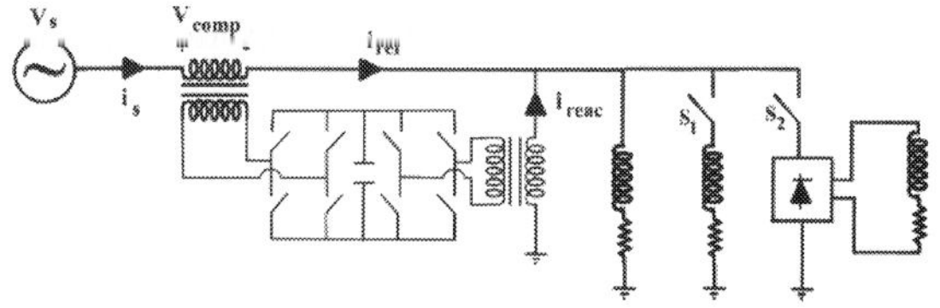

Figure 9. General test system circuit

This power system consists of a three phase 380V (RMS, L-L), 50 Hz utility, two three phase linear R-L load and a three phase rectifier as a nonlinear load which can be connected to the circuit at different times. This is for the investigation of the proposed control system capability in dynamic conditions. For the investigation of the voltage sag and swell conditions, utility voltages have 0.25 percent sag between 0.04 sec and 0.08 sec and 0.25 percent swell between 0.08 sec and 0.12 sec. Also, for the investigation of the proposed control strategy in the harmonic conditions, source voltage has been harmonized between

0.17 sec and 0.4 sec. Table 1, shows the utility voltage data and Table 2, shows the load powers and related switching times. In this study series active filter has been connected to the circuit at time zero. But parallel active filter has been connected to the circuit at time 0.25 sec. A number of selected simulation results will be shown later.

Voltage Order	Magnitude (pu)	Phase Angle (deg)	Time (sec)
1	1	0	0-0.04, 0.12-0.4
1	0.75	0	0.04-0.08
1	1.25	0	0.08-0.12
5	0.1	-45	0.17-0.4
3	0.1	0	0.17-0.4

Table 1. Utility voltage harmonic and sequence parameters data

Load	Nominal Power (kVA)	Nominal Voltage (RMS, L-L)	Switching Time (Sec)
Linear	10	380V	0
Linear	10	380V	0.29
Non linear	5	380V	0.33

Table 2. Load powers and related switching times data

Figure 10, shows the source side voltage of phase 1. Figure 11 shows the compensated load side voltage of phase 1. Figure 12 shows SAF voltage of phase 1. Figure 13 and 14 shows first order component magnitude of the source voltage extracted by T and T/2 data windows respectively. Figure 15 shows first order component magnitude of the source voltage extracted by T/12 data window. Figure 16 shows the first order component magnitude of the source voltage extracted by the proposed composition of T/2 and T/12 data windows. Figure 17 shows the load side current of phase 1. Figure 18 shows the source side current of phase

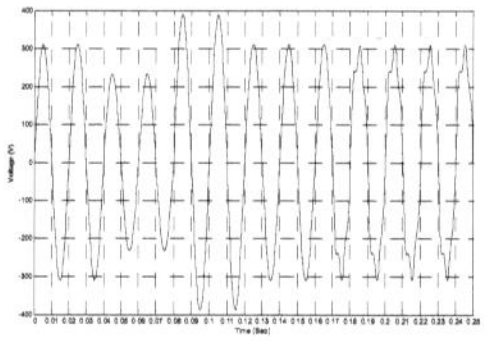

Figure 10. Source side voltage of phase 1 (sag has been occurred between 0.04 sec and 0.08 sec and swell has been occurred between 0.08 sec and 0.12 sec. Also, harmonics have been concluded between 0.17 sec to 0.4 sec)

1. Figure 19 shows the PAF current of phase 1. Figure 20 and 21 show first order component magnitude of the load current extracted by T and T/2 data windows respectively. Figure 22 shows the first order component magnitude of the load current extracted by T/12 data window. Figure 23 shows the first order component magnitude of the load current extracted

by the proposed composition of T/2 and T/12 data windows. Figure 24 and 25 show the source side and load side voltages frequency spectrum, respectively. Finally Figures 26 and 27 show the load side and source side currents frequency spectrum, respectively. Table 3 shows THDs of the source and load voltages and currents. Load voltage and source current harmonics have been compensated satisfactory.

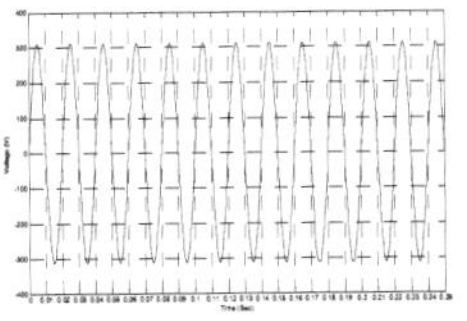

Figure 11. Load side voltage (sag and swell as well as harmonics have been compensated)

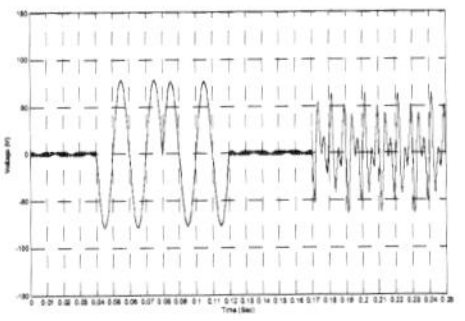

Figure 12. Compensator voltage (this is only between sag, swell, and harmonic times)

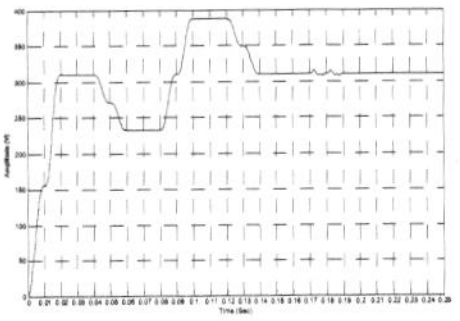

Figure 13. Extracted source side voltage magnitude (in this state data window length is *T* and settling time is 0.02 sec)

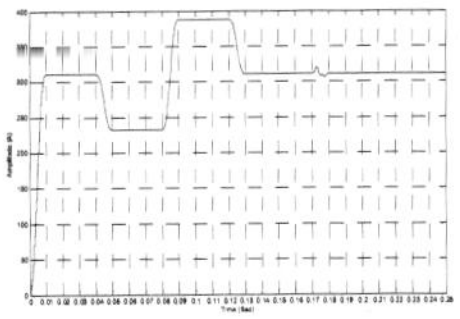

Figure 14. Extracted source side voltage magnitude (in this state data window length is T/2 and settling time is 0.01 sec)

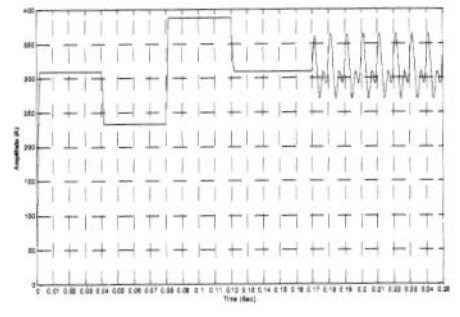

Figure 15. Extracted source side voltage magnitude by T/12 data window (in this state settling time is 0.00166 sec but, there are unfiltered oscillations in the response)

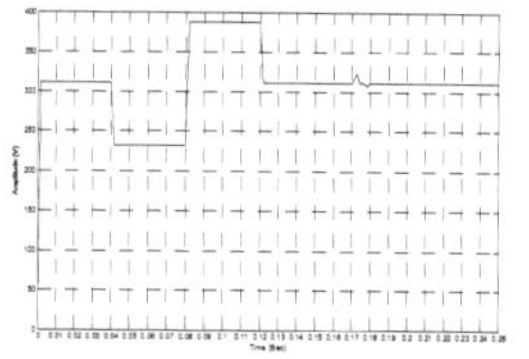

Figure 16. Extracted source side voltage magnitude by the proposed approach (in this state composition of T/2 and T/12 data windows have been used, settling time is 0.00166 sec and there is no oscillation in the response)

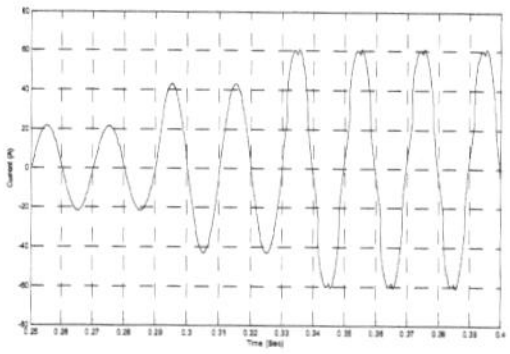

Figure 17. Load side current (there is a linear three phase load until 0.29 sec, another linear three phase load has been connected to the circuit at time 0.29 sec and finally a nonlinear rectifier load has been connected to the circuit at time 0.33 sec)

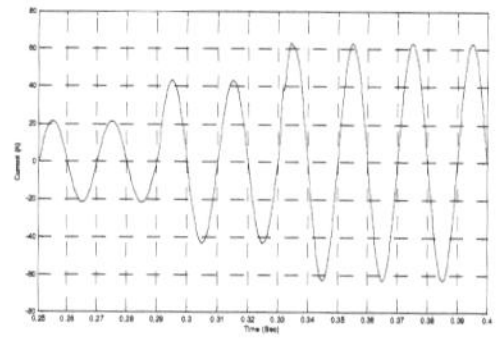

Figure 18. Source side current (load current harmonics as well as reactive power have been compensated)

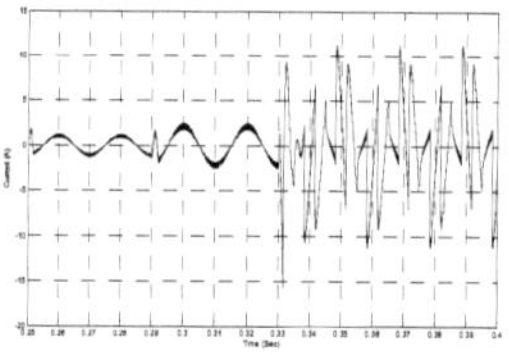

Figure 19. Compensator current (it is for reactive power compensation as well as current harmonics)

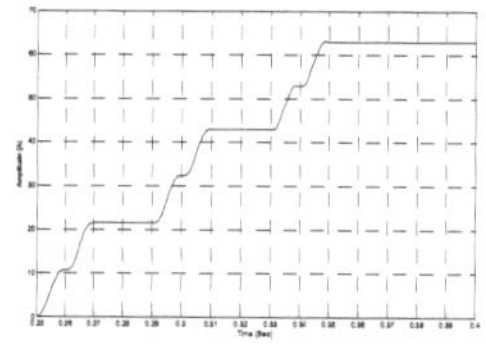

Figure 20. Extracted load side current magnitude (in this state data window length is *T* and settling time is 0.02 sec)

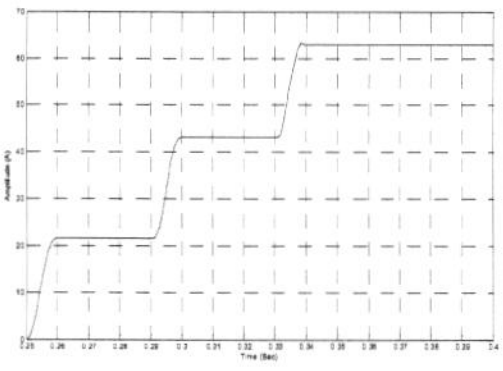

Figure 21. Extracted load side current magnitude (in this state data window length is T/2 and settling time is 0.01 sec)

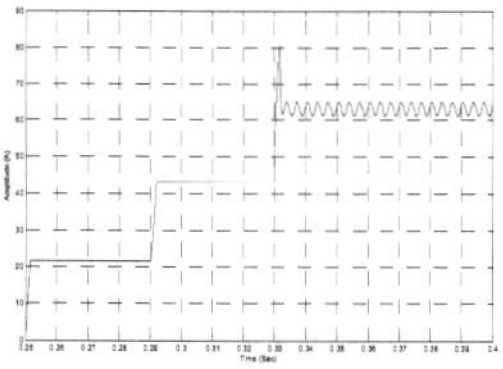

Figure 22. Extracted load side current magnitude by T/12 data window (in this state settling time is 0.00166 sec but there are unfiltered oscillations in the response)

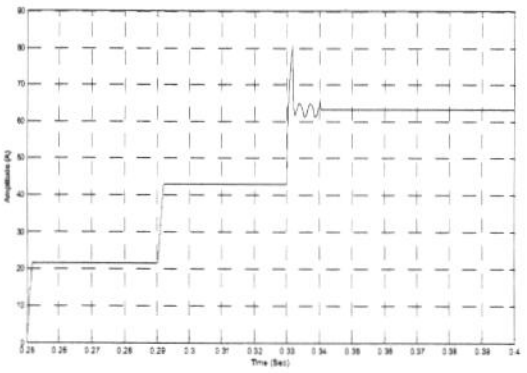

Figure 23. Extracted load side current magnitude by the proposed approach (in this state composition of T/2 and T/12 data windows have been used, settling time is 0.00166 sec and there is no oscillation in the response)

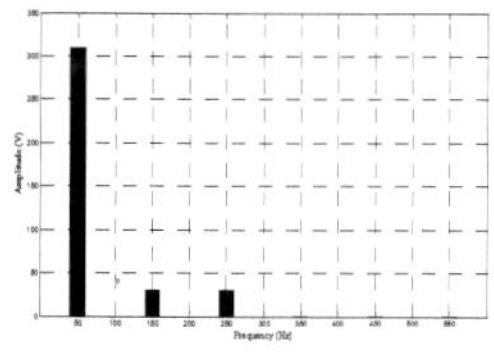

Figure 24. Source voltage harmonic spectrum (it has third and fifth harmonics)

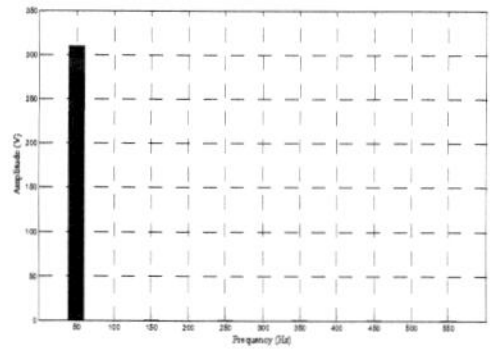

Figure 25. Load voltage harmonic spectrum (harmonics have been compensated)

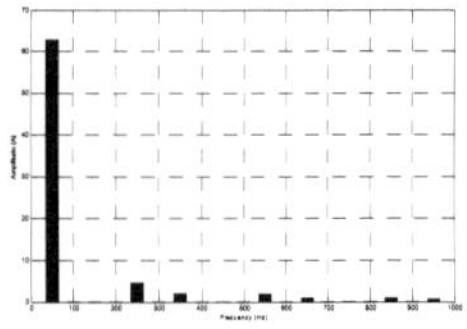

Figure 26. Load side current harmonic spectrum

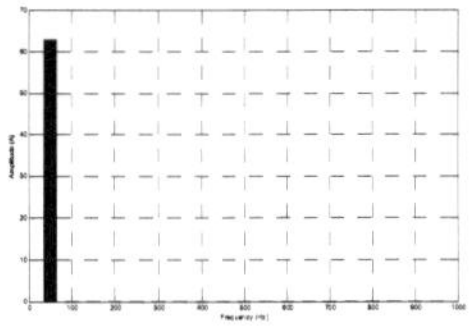

Figure 27. Source side current harmonic spectrum

Vs THD	IL THD	VL THD	Is THD
0.14	0.09	.0002	.0001

Table 3. Total Harmonic Distortion (THD)

7.2. Results of MOADALINE approach

For the investigation of the validity of the mentioned control strategy of MOADALINE for power quality compensation of a distribution system, simulation of the test circuit has been done in MATLAB software. Source current and load voltage have been measured and analyzed in the proposed control system for the determination of the compensator signals of SAF and PAF. Related equations of the controlled system have been compiled in MATLAB software via M-file. Desired values of a_n and b_n are determined in the proposed algorithm for extraction of the reference signals.

The power system consists of a harmonized and unbalanced three phase 380V (RMS, L-L), 50 Hz utility, a three phase rectifier as a nonlinear load, a three phase balanced R-L load which is connected to the circuit at 0.04 sec and a one phase load which is connected to the circuit at 0.07 sec.

For the investigation of the voltage harmonic condition, utility voltages have harmonic and negative sequence components between 0.03 sec and 0.1 sec. Also, for the investigation of the proposed control strategy in unbalance condition, magnitude of the first phase voltage is increased to the 1.25 pu between 0.02 sec and 0.04 sec and decreased to the 0.75 pu between 0.06 sec to 0.08 sec. Figure 28 shows the source side voltage of phase 1. Figure 29, shows the compensator voltage of phase 1. Figure 30, shows load side voltage of phase 1. Figure 31, shows the load side current of phase 1. Figure 32, shows the reactive current of phase 1. Figure 33, shows the harmonic current of phase 1. Finally Figure 34, shows the source side current of phase 1. Load voltage and source current harmonics have been compensated satisfactory.

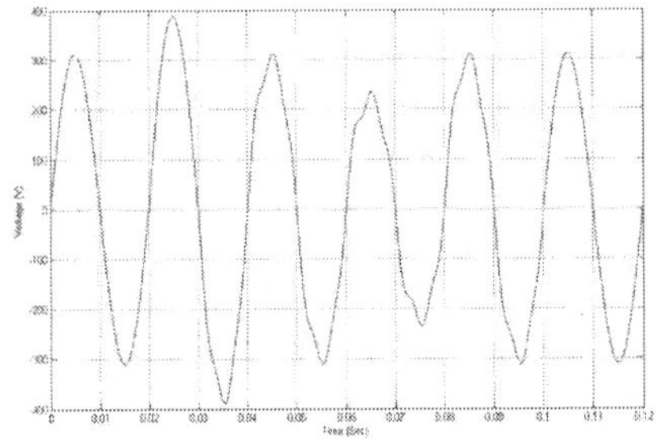

Figure 28. Source side voltage of phase 1

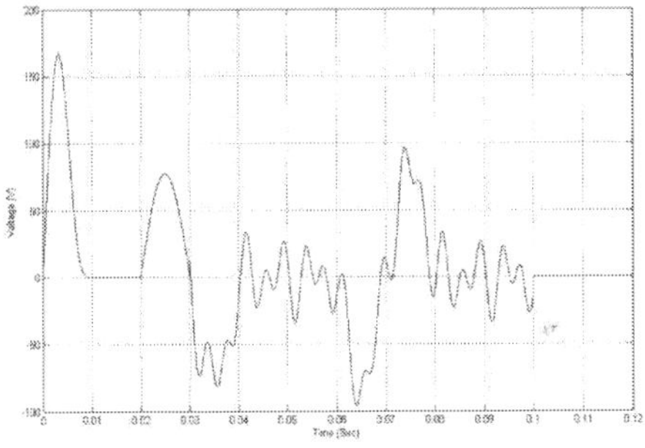

Figure 29. Compensator voltage of phase 1

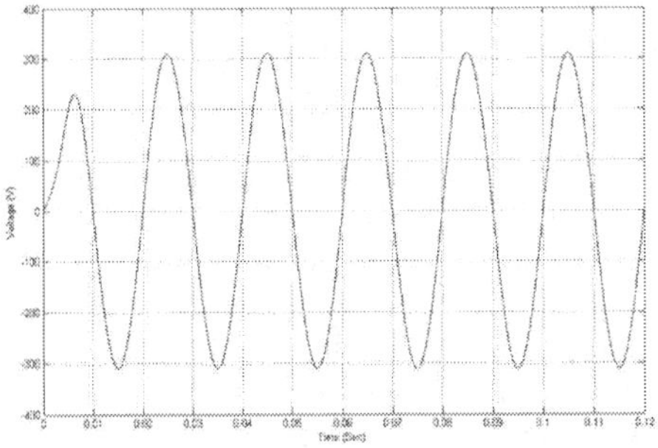

Figure 30. Load side voltage of phase 1

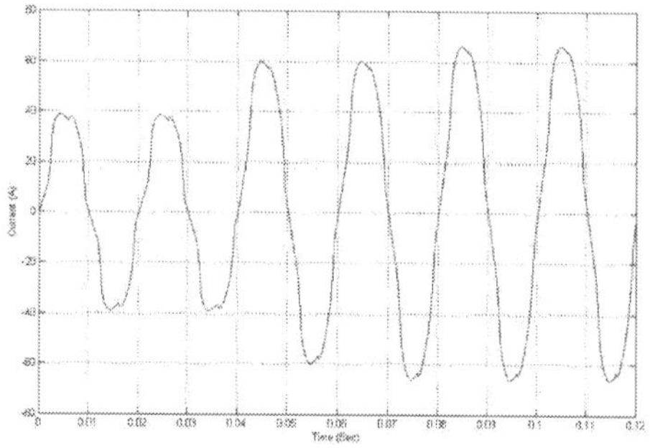

Figure 31. Load side current of phase 1

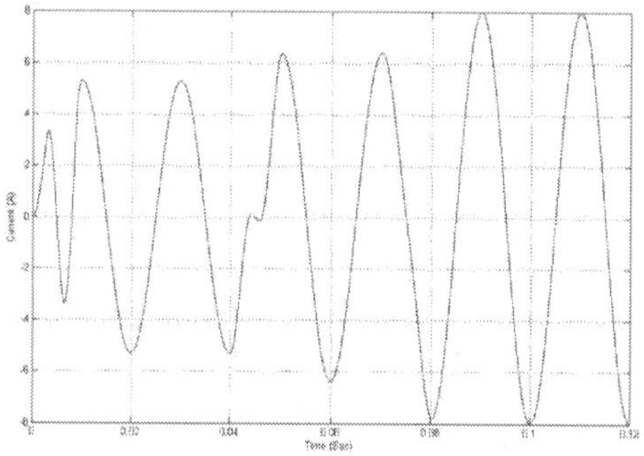

Figure 32. Reactive current of phase 1

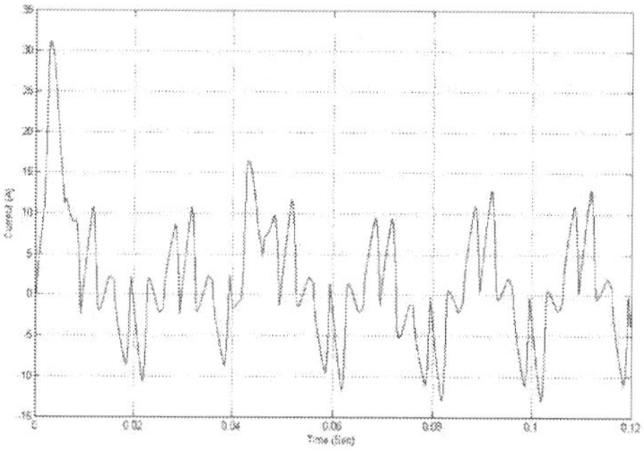

Figure 33. Harmonic current of phase 1

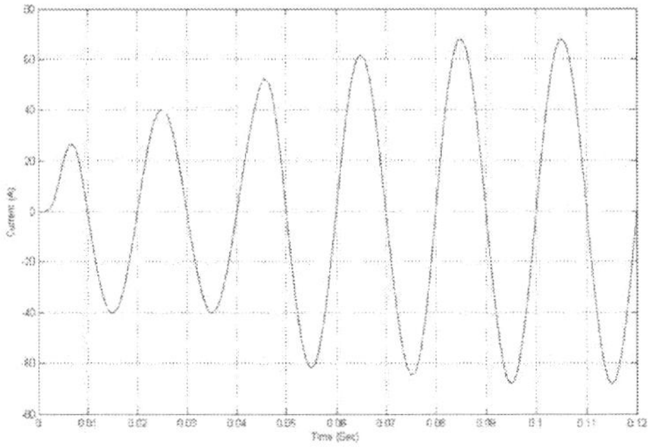

Figure 34. Source side current of phase 1

7.3. Results of power flow control

For the investigation of the validity of the power flow control strategy in a distribution system, simulation of the test circuit of Figure 35 has been done.

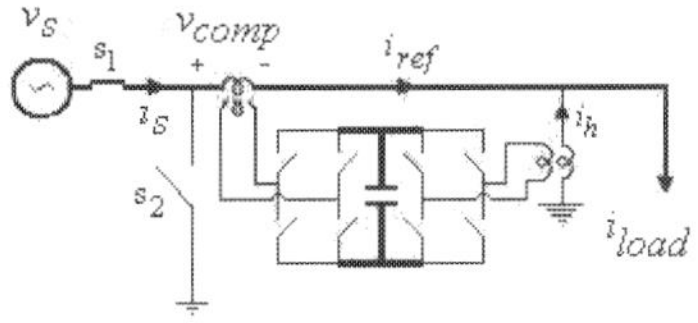

Figure 35. General test system circuit

This power system consists of a harmonized and unbalanced three phase 20 kV (RMS, L-L), 50 Hz utility, a three phase balanced R-L load and a nonlinear three phase load. For the investigation of the voltage harmonic condition, utility voltages have harmonic and negative sequence components between 0.15 s and 0.35 s. Also, for the investigation of the proposed control strategy in unbalance condition, magnitude of the first phase voltage is increased to the 1.25 pu between 0.10 s and 0.20 s and decreased to the 0.75 pu between 0.3 s to 0.4 s. Investigation of the control circuit performance in fault condition is done by a three phase fault in output terminal of the main source between 0.4 s to 0.5 s. Table 4 shows the utility voltage harmonic and sequence parameters data and Table 5 shows the load power and voltage parameters. Table 6 show states of switches of s_1 and s_2. A number of selected simulation results have been shown next.

Voltage order	Sequence	Magnitude (pu)	Phase angle (degree)	Time duration (sec)
5	+	0.12	-45	0.15-0.35
3	-	0.1	0	0.15-0.35

Table 4. Utility voltage harmonic and sequence parameters data

Load	Nominal power (kVA)	Nominal voltage (RMS, L-L)
Linear	50	20 kV
Non linear	260	20 kV

Table 5. Load power and voltage parameters data

switch	0<t<0.4 sec		0.4<t<0.5 sec	
	First strategy	Second strategy	First strategy	Second strategy
s_1	close	close	open	open
s_2	open	open	close	close

Table 6. States of switches

Figure 36, shows the source side voltage of phase 1. Figure 37, shows the compensator voltage of phase 1. Figure 38, shows the load side voltage of phase 1. Figure 39, shows the load side current of phase 1. Figure 40, shows the compensator current of phase 1. Figure 41, shows the source side current of phase 1. Figure 42, shows the load active and reactive powers. Figure 43, shows generation PAF active and reactive powers. Figure 44, shows the consumption active and reactive powers of the series active filter. Figure 45, shows generation active and reactive powers of the source. Figure 46, shows the difference between load voltage magnitude and nominal value and finally Figure 47, shows the difference between load voltage and source voltage phases.

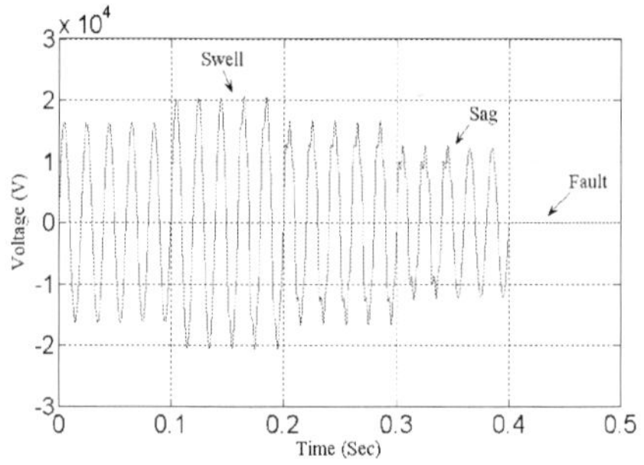

Figure 36. Source side voltage of phase 1 (a swell has been occurred between 0.1 and 0.2 sec and a sag has been occurred between 0.3 and 0.4 sec. Also positive and negative harmonic sequences have been concluded between 0.15 and 0.35 sec. It has been tripped between 0.4 and 0.5 sec)

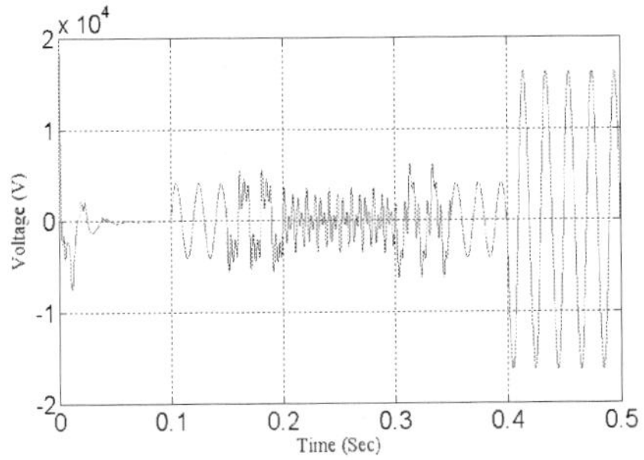

Figure 37. Compensator voltage of phase 1 (compensator voltage has been determined for the sag, swell, interruption, negative sequence and harmonics improvement)

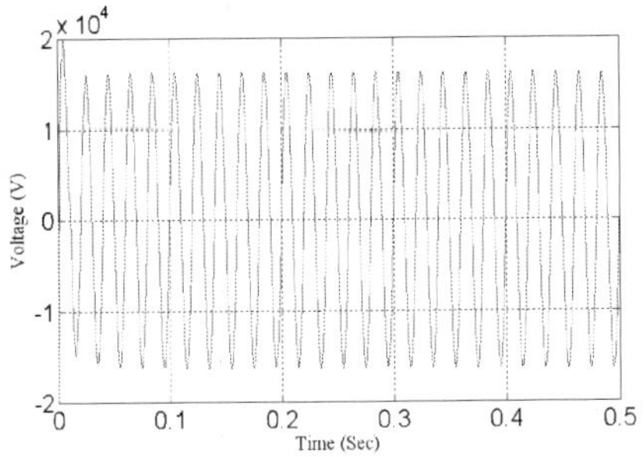

Figure 38. Load side voltage of phase 1 (swell, sag, positive and negative harmonic sequences have been improved. Load voltage has been compensated in the fault condition)

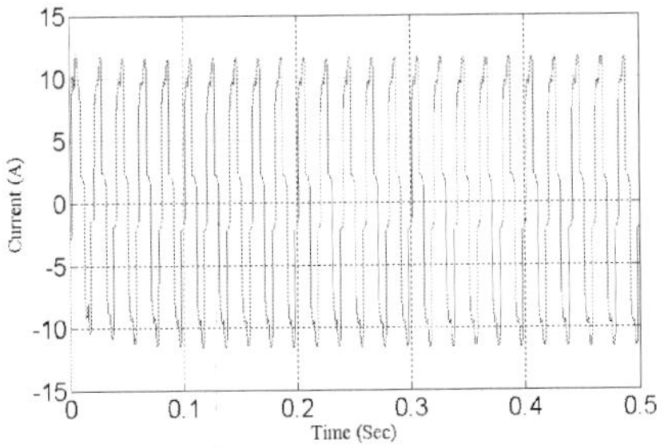

Figure 39. Load side current of phase 1 (it has different order harmonics. It should be considered that this current has been calculated after the voltage compensation, so the voltage unbalance has not been concluded in the current)

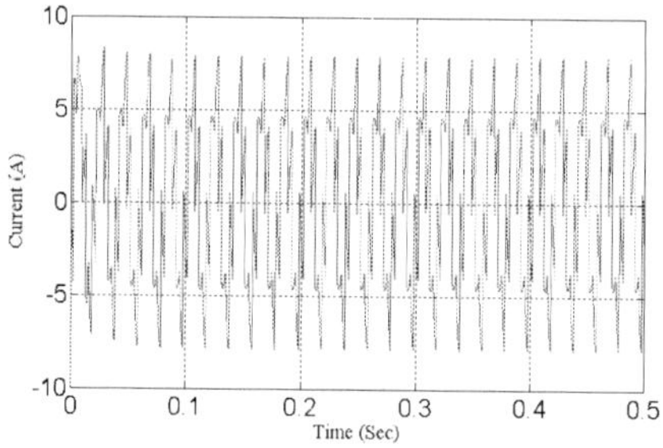

Figure 40. Compensator current of phase 1 (compensator current has been determined for the load current harmonics improvement as well as the reactive power)

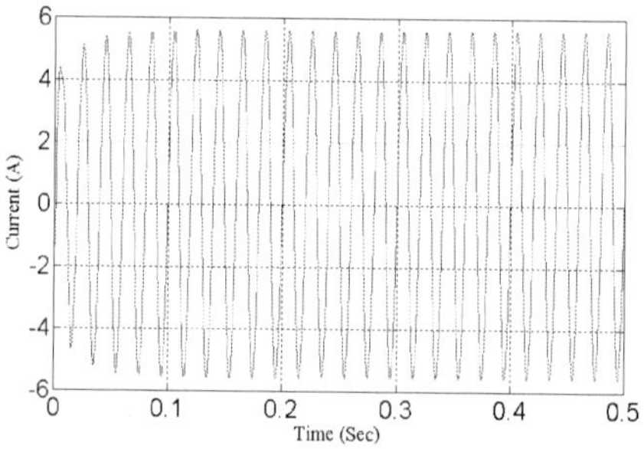

Figure 41. Source side current of phase 1 (harmonics and reactive power components of the load current have been canceled)

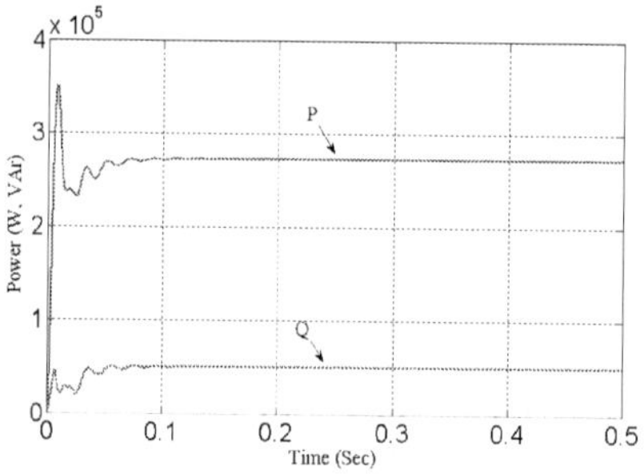

Figure 42. Load active and reactive power (load is nonlinear resistive-inductive)

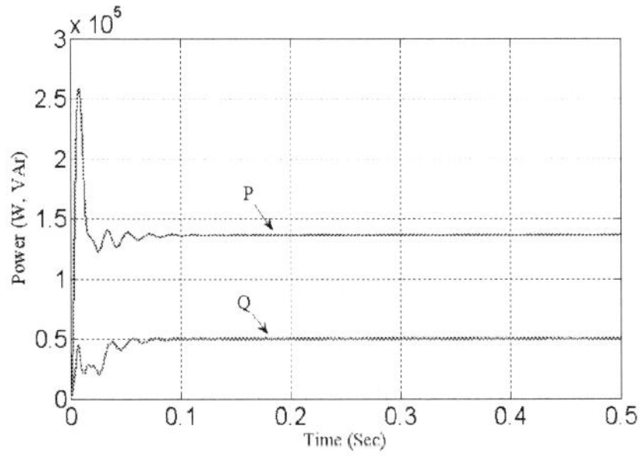

Figure 43. Generation PAF active and reactive powers (reactive power and harmonics of load current have been supplied by the parallel active filter)

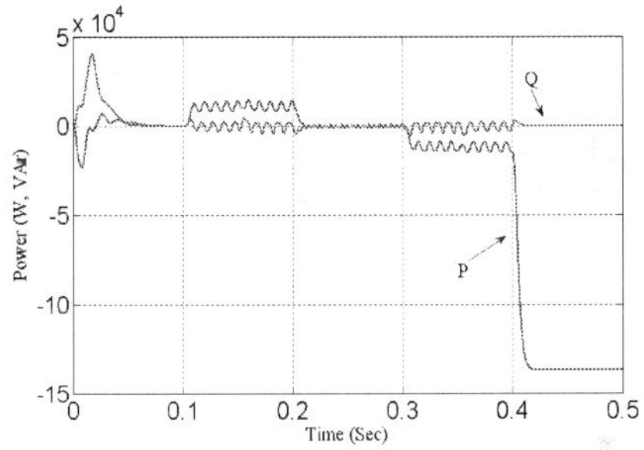

Figure 44. Consumption active and reactive power of series active filter (it is considered that the consumption power has been increased between sag times and decreased between swell times. Cause of power oscillation has been investigated in the text. Active power has been increased in fault condition for prevention of load interruption)

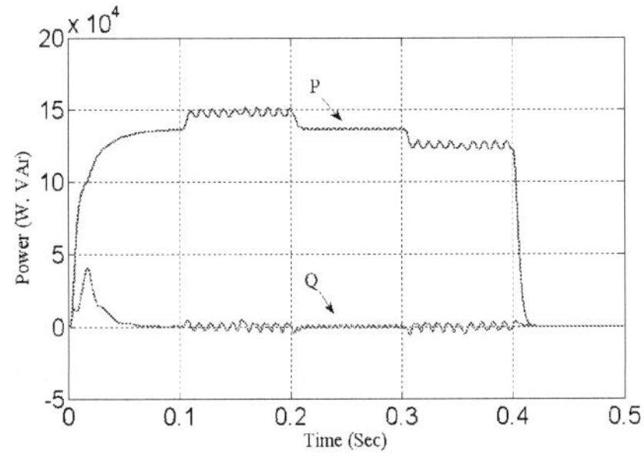

Figure 45. Generation active and reactive power of source (it is considered that the generation active power has been increased between sag times and decreased between swell times. Reactive power has been compensated. Cause of power oscillation has been investigated in the text. These powers are equal to zero in fault condition)

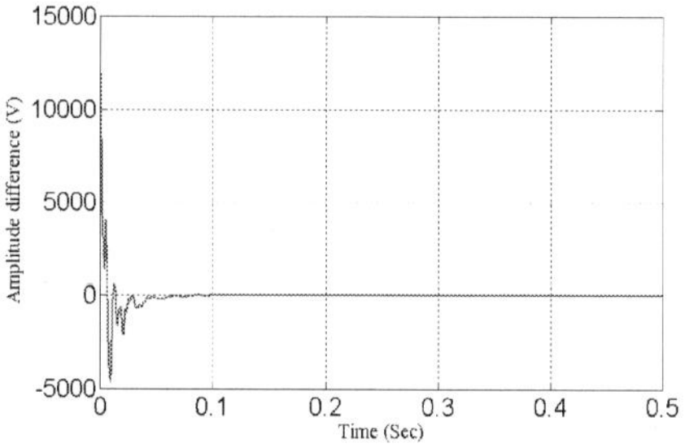

Figure 46. Difference between load voltage magnitude and nominal value (this amount is decreased by use of the PI controller to zero)

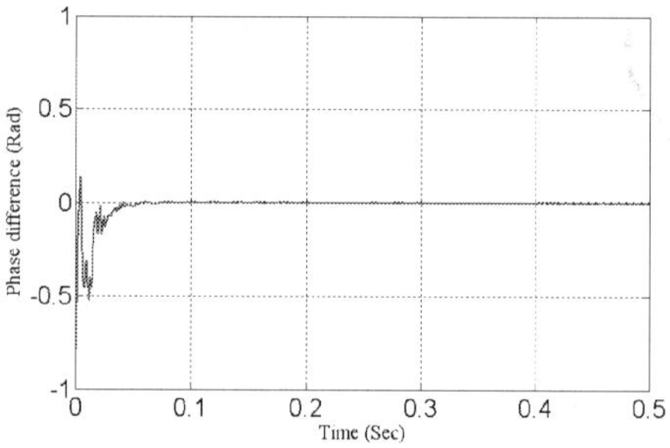

Figure 47. Difference between load voltage and source voltage phases (this amount decreased by use of PI controller to zero)

8. Conclusions

In this research different approaches for reference generation of UPQC have been proposed. Based on the general equations of Fourier transform, its response settling time is one cycle. In this research for increasing the response speed and improving the control system capabilities in dynamic conditions, very fast Fourier transform approach has been proposed for balanced three phase systems. In the proposed approach, settling time of the response could be reduced to one twelfth of a cycle. In the proposed approach, there were two data window lengths, T/12 and T/2. In the sag, swell, and load change conditions, control system was switched to T/12 for obtaining fast response. Then for improving the proposed approach responsibility in filtering of unwanted steady state oscillation, control system was switched to T/2 for obtaining no oscillated response. In these states, fast response in dynamic conditions as well as good response in the steady state conditions would be possible. In this research for the detection of the source voltage sag, swell, and load change conditions, derivative of the first order magnitude of the voltage and current signal, were compared to a constant value. This was based on the fact that in these conditions voltage or current magnitude generally changes rapidly. Proposed control approach was simulated in MATLAB/SIMULINK software. Voltage sag, swell, and harmonics were compensated by

SAF. But, reactive power and current harmonics were compensated by PAF. THD of load voltage before compensation was 14.14 percent which was reduced to almost zero after the compensation. But, THD of the source current before compensation was 9 percent which was reduced to almost zero after the compensation.

Also the proposed reference generation algorithm based on MOADALINE has been compiled in MATLAB software via M-File. Voltage harmonics have been compensated by SAF of the UPQC and current harmonics have been compensated by PAF of the UPQC. Based on the results proposed strategy not only could generate pure sinusoidal source current and load voltage but also could compensate source reactive power satisfactory. Total harmonic distortion of load voltage and current before compensation was 0.17 and 0.12 respectively which was reduced to almost zero after the compensation.

Another scope of this research was reference generation based on power flow control. This approach was based on relation between active and reactive powers and load voltage. In this approach amount of reactive power arranged to zero but amount of active power arranged to load nominal power. Also a PI controller used for arranging the load voltage magnitude to the nominal amount.

Author details

Ahad Mokhtarpour
Department of Electrical Engineering, Tabriz Branch, Islamic Azad University, Tabriz, Iran

Heidarali Shayanfar
Department of Electrical Engineering, South Tehran Branch, Islamic Azad University, Tehran, Iran

Seiied Mohammad Taghi Bathaee
Department of Electrical Engineering, K.N.T University, Tehran, Iran

9. References

[1] Fujita H., Akagi H. The Unified Power Quality Conditioner: The Integration of Series and Shunt Active Filters. IEEE Transaction on Power Electronics 1998; 13(2) 315-322.

[2] Hannan M. A., Mohamed A. PSCAD/EMTDC Simulation of Unified Series-Shunt Compensator for Power Quality Improvement. IEEE Transaction on Power Delivery 2005; 20(2) 1650-1656.

[3] Shayanfar H. A., Mokhtarpour A. Management, Control and Automation of Power Quality Improvement. In: Eberhard A. (ed.) Power Quality. Austria: InTech; 2010. p127-152.

[4] Khadkikar V., Chandra A. A Novel Structure for Three Phase Four Wire Distribution System Utilizing Unified Power Quality Conditioner (UPQC). IEEE Transactions on Industry Applications 2009; 45(5) 1897-1902.

[5] Kwan K.H., Chu Y.C., So P.L. Model-Based H∞ Control of a Unified Power Quality Conditioner. IEEE Transactions on Industrial Electronics 2009; 56 (7) 2493-2504.

[6] Khadkikar V., Chandra A. A New Control Philosophy for a Unified Power Quality Conditioner (UPQC) to Coordinate Load-Reactive Power Demand Between Shunt and Series Inverters. IEEE Transactions on Power Delivery 2008; 23 (4) 2522-2534.

[7] Lee W.C., Lee D.M., Lee T.K. New Control Scheme for a Unified Power Quality Compensator-Q with Minimum Active Power Injection. IEEE Transactions on Power Delivery 2010; 25(2) 1068-1076.

[8] Han B., Bae B., Kim H., Baek S. Combined Operation of Unified Power-Quality Conditioner with Distributed Generation. IEEE Transaction on Power Delivery 2006; 21(1) 330-338.

[9] Mohammadi H.R., Varjani A.Y., Mokhtari H. Multiconverter Unified Power-Quality Conditioning System: MC-UPQC. IEEE Transactions on Power Delivery 2009; 24(3) 1679-1686.

[10] Mokhtarpour A., Shayanfar H.A., Bathaee. S.M.T Extension of Fourier Transform for Very Fast Reference Generation of UPQC. International Journal on Technical and Physical Problems of Engineering 2011; 3(4) 120-126.

[11] Mokhtarpour A., Shayanfar H.A., Bathaee. S.M.T UPQC Control Based on MO-ADALINE Approach. International Journal on Technical and Physical Problems of Engineering 2011; 3(4) 115-119.

Section 3

Power Quality Improvement in End Users Stage

Chapter 7

Power Quality Improvement Using Switch Mode Regulator

Raju Ahmed and Mohammad Jahangir Alam

Additional information is available at the end of the chapter

http://dx.doi.org/10.5772/53234

1. Introduction

Power quality describes the quality of voltage and current. It is an important consideration in industries and commercial applications. Power quality problems commonly faced are transients, sags, swells, surges, outages, harmonics and impulses [1]. Among these voltage sags and extended under voltages have large negative impact on industrial productivity, and could be the most important type of power quality variation for many industrial and commercial customers [1-5].

Voltage sags is mainly due to the fault occurring in the transmission and distribution system, loads like welding and operation of building construction equipment, switching of the loaded feeders or equipments. Both momentary and continuous voltage sags are undesirable in complex process controls and household appliances as they use precision electronic and computerized control.

Major problems associated with the unregulated long term voltage sags include equipment failure, overheating and complete shutdown. Tap changing transformers with silicon-controlled rectifiers (SCR) are usually used as a solution of continuous voltage sags [6]. They require large transformer with many SCRs to control the voltage at the load which lacks the facility of adjusting to momentary changes. Some solutions have been suggested in the past to encounter the problems of voltage sag [7-11]. But these proposals have not been realized practically to replace conventional tap changing transformers.

Now a day's various power semiconductor devices are used to raise power quality levels to meet the requirements [12]. Several AC voltage regulators have been studied as a solution of voltage sags [13-18]. In [13] the input current was not sinusoidal, in [14-16] the efficiency of the regulator was not analyzed and in [17-18] the input power factor was very low and the efficiency is also found poor. Compact and fully electronic voltage regulators are still unavailable practically.

Dynamic Voltage Restorer (DVR) is sometimes used to regulate the load side voltage [19-21]. The DVR requires energy storage device to compensate the voltage sags. Flywheels, batteries, superconducting magnetic energy storage (SMES) and super capacitors are generally used as energy storage devices. The rated power operation of DVR depends on the size and capacity of energy storage device which limits its use in high power applications. Whereas, switching regulator needs no energy storage devices, therefore, can be used both in low power and high power applications.

The objective of this chapter is to describe the operation and design procedure of a switch mode AC voltage regulator. Firstly, some reviews of the regulators are presented then the procedure of design and analysis of a switch mode regulator is described step by step. Simulation software OrCAD version 9.1 [22] is used to analyze the regulator. The proposed regulator consists mainly two parts, power circuit and control circuit. The power circuit consist two bi-directional switches which serve as the freewheeling path for each other. A signal generating control circuit is to be associated with the power circuit for getting pulses of the switches. In the control circuit, a commercially available pulse width modulator IC chip SG1524B is used, thus circuit is compact and more viable.

2. Review of voltage regulators

2.1. Switching-mode power supply (SMPS)

A switching-mode power supply (SMPS) is switched at very high frequency. Conversion of both step down and step up of voltage is possible using SMPS. Uses of SMPSs are now universal in space power applications, computers, TV and industrial units. SMPSs are used in DC-DC, AC-AC, AC-DC, DC-AC conversion for their light weight, high efficiency and isolated multiple outputs with voltage regulation. Main parts of a Switching-mode power supply are:

(a) Power circuit, (b) Control circuit.

Figure 1 shows the block diagram of a SMPS. The power circuit is mainly the input, output side with the switching device. The switching device is continuously switched at high frequency by the gate signal from the control circuit to transfer power from input to the output. The control circuit of a SMPS basically generates high frequency gating pulses for the switching devices to control the output voltage. Switching is performed in multiple pulse width modulation (PWM) fashion according to feedback error signal from the load. High frequency switching reduces filter requirements at the input and output sides of the converter. Simplest PWM control uses multiple pulse modulations generated by comparing a DC with a high frequency carrier triangular wave.

The PWM control circuit is commonly available as integrated form. The designer can select the switching frequency by choosing the value of RC to set oscillator frequency. As a rule of thumb to maximize the efficiency, the oscillation period should be about 100 times longer than the switching time of the switching device such as Transistor, Metal oxide semiconductor field-effect transistor (MOSFET), Insulated gate bipolar transistor (IGBT). For example, if a switch has a switching time of 0.5 us, the oscillator period would be 50 us, which gives the maximum oscillation frequency of 20 KHz. This limitation is due to the switching loss in the switching devices. The switching loss of switching devices increases with the switching frequency. In addition, the core loss of inductor limits the high frequency operation.

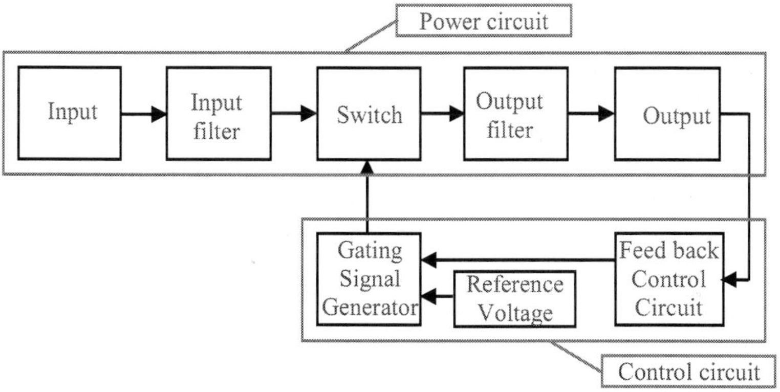

Figure 1. Block diagram of Switching-mode power supply (SMPS).

2.2. DC-DC converter

Figure 2 illustrates the circuit of a classical linear power converter. Here power is controlled by a series linear element; either a resister or a transistor is used in the linear mode. The total load current passes through the series linear element. In this circuit greater the difference between the input and the output voltage, more is the power loss in the controlling device (linear element). Linear power conversion is dissipative and hence is inefficient.

The circuit of Fig. 3 illustrates basic principle of a DC-DC switching-mode power converter. The controlling device is a switch. By controlling the duty cycle, (the ratio of the time in on positions to the total time of on and off position of a switch) the power flow to the load can be controlled in a very efficient way. Ideally this method is 100% efficient. In practice, the efficiency is reduced as the switch is non-ideal and losses occur in power circuits. Hence, one of the prime objectives in switch mode power conversion is to realize conversion with the least number of components having better efficiency and reliability. The DC output voltage to the load can be controlled by controlling the duty cycle of the rectangular wave supplied to the base or gate of the switching device. When the switch is on, it has only a small saturation voltage drop across it. In the off condition the current through the switch is zero.

The output of the switch mode power conversion circuit (Fig. 3) is not pure DC. This type of output is applicable in some cases such as oven heating without proper filtration. If constant DC is required, then output of converter has to be smoothed out by the addition of low-pass filter.

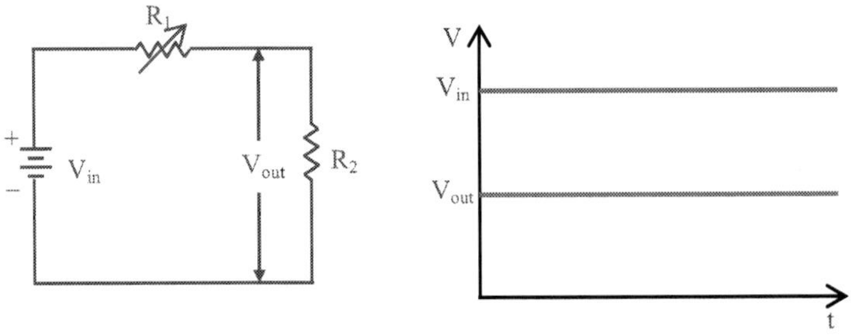

Figure 2. Linear (dissipative) DC-DC power conversion circuit.

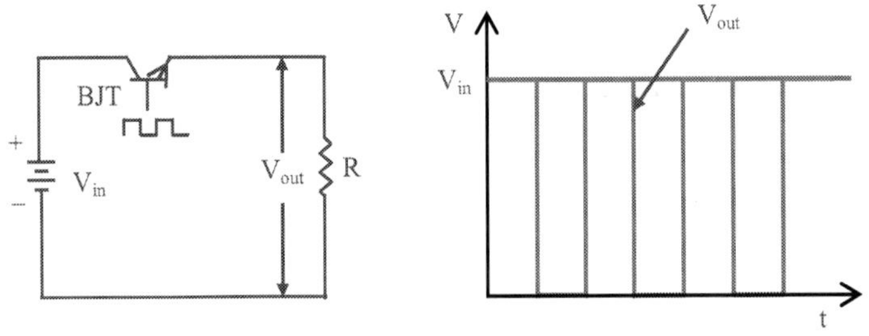

Figure 3. Switching-mode (non dissipative) DC-DC power conversion circuit.

2.2.1. Types of DC-DC converter

There are four basic topologies of switching DC-DC regulators:

a. Buck regulator
b. Boost regulator
c. Buck-Boost regulator and
d. Cûk regulator.

The Circuit diagram of four basic DC-DC switching regulators is shown in Fig. 4. The expression of output voltage for the four types of DC-DC regulators are as follows:

For Buck regulator, $V_{out} = kV_{in}$, For Boost regulator, $V_{out} = \frac{V_{in}}{1-k}$

For Buck- Boost regulator and Cûk regulator, $V_{out} = \frac{-kV_{in}}{1-k}$

Where k is the duty cycle, the value of k is less than 1. For Buck regulator output voltage is always lower than input voltage, for Boost regulator output voltage is always higher than input voltage. For Buck-Boost regulator and Cûk regulator output voltage is higher than input voltage when the value of k is higher than 0.5, and output voltage is lower than input voltage when the value of k is lower than 0.5. When k is equal to 0.5 output voltage is same as input voltage. In Buck-Boost and Cûk regulator, the polarity of output voltage is opposite to that of the input voltage, therefore theses regulators are also called inverting regulators.

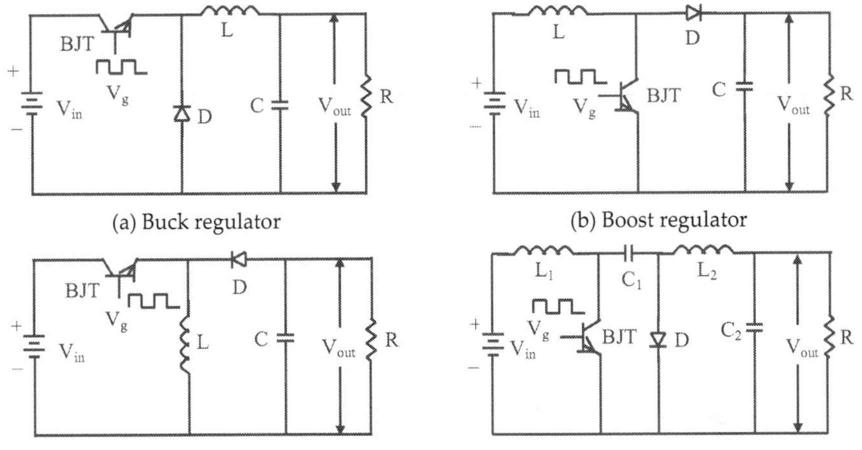

(a) Buck regulator (b) Boost regulator (c) Buck-Boost regulator (d) Cûk regulator

Figure 4. Circuit diagram of DC-DC regulator, (a) Buck regulator, (b) Boost regulator, (c) Buck-Boost regulator, (d) Cûk regulator.

2.3. AC-AC converter

The AC voltage regulator is an appliance by which the AC output voltage can be set to a desired value and can be maintained constant all the time irrespective of the variations of input voltage and load. This subject is vast and the field of application extends from very large power systems to small electronic apparatus. Naturally, the types of regulators are also numerous. The design of the regulators depends mainly on the power requirements and degree of stability.

The AC voltage can be regulated by the following ways.

a. Solid-state tap changer and steeples control by variac
b. Solid-tap changer using anti-parallel SCRs
c. Voltage regulation using servo system
d. Phase controlled AC regulator
e. Ferro-resonant AC regulator
f. Switch mode AC regulator

2.3.1. Solid-state tap changer and stepless control by variac

The voltage regulations by tap-changing switches are used in many industrial applications where the maintenance of output voltage at a constant value is not very stringent, such as ordinary battery chargers, electroplating rectifiers etc. For smaller installation, off-load tap changing switches are used and for large installation on-load tap changing switches are used. The switches are generally incorporated at the secondary of the transformer. For a low voltage high current load, the switches are provided on the primary side of the transformer due to economical reason. For line voltage correction, taps are provided on the primary of the transformer. For three-phase transformers three pole tap changing switches are used.

In off-load tap changer, the output is momentarily cut-off from the supply. It is therefore used for low capacity equipment and where the momentary cut-off of the supply is not objectionable for the load. The major limitation of the off-load tap changing switches is the occurrences of arcs at the contact points during the change-over operation. This shortens the life of off-load rotary switches, particularly of high current ratings. In Fig. 5 (a), three four-position switches of an off load tap changer are shown, such that the minimum of X volts per step are available at the output.

The voltage is corrected by tap-charging switches in steps. Where stepless control is required, variable autotransformers or variacs are used. The normal variac consists of a toroidal coil wound on a laminated iron ring. The insulation of the wire is removed from one of the end faces and the wire is grounded to ensure a smooth path for the carbon brush. Carbon brush is used to limit the circulating current, which flows between the short-circuited turns.

A Buck-Boost transformer is sometimes used for AC voltage regulation when the output voltage is approximately the same as the mean input voltage as shown in Fig. 5(b). In this case if the output voltage is less than or greater than the desired value, it can be increased or decreased to the desired value by adding a suitable forward or reverse voltage with the input through the Buck-Boost transformer.

(a) Off load tap changing switch arrangement.

(b) Voltage control by combination of a Buck–Boost transformer and a variac.

Figure 5. Circuit diagram of AC voltage controller using (a) Off load tap changer and (b) Buck-Boost transformer and variac.

2.3.2. Solid tap changer using anti-parallel SCRs

Anti-parallel SCRs combinations can replace the voltage sensitive relay in the tap-changing regulator. Figure 6 shows a tap changer with three taps which can be connected to the load through three anti-parallel switches. When the SCR1-SCR2 switch is fired, tap 1 is connected to the load. Similarly taps 2 and 3 can be connected to the load through the SCR3-SCR4 and CSR5-SCR6 switches respectively. Thus, any number of taps can be connected to the load with similar SCR switches. When one group of SCRs operates for the whole cycle and other groups are off, the voltage corresponding to the tap of that group appears at the load. Changeover from one loop to the other is done simply by shifting the firing pulses from one group of SCRs to the other.

With resistive load, the load current becomes zero and the SCRs stop conduction as soon as the voltage reverses its polarity. Therefore, when one group is fired, the other groups are commutated automatically. With reactive loads, the situation is complicated by the fact that the zero current angle depends on the load power factor. This means that the SCR conducts a finite value of current at the time of reversal of line voltage. This results in either preventing a tap change due to reverse bias on the SCR to be triggered or causing a short circuit between the taps through two SCRs.

2.3.3. *Voltage regulation using servo system*

Voltage regulators using servo systems are quite common. Both single and three-phase types are available. The rating of this type of regulator is quite high and is more economical for high power rating. This regulator normally consist a variac driven by a servomotor, a sensing unit and a voltage and power amplifier to drive the motor in a reversible way. Various types of driving motor may be used for regulating the unit, such as direct current, induction and synchronous motors. However, in all cases, the motor must come to rest rapidly to avoid overrun and hunting. The amount of overrun may be reduced by dynamic braking in the case of a DC motor or by disconnecting the motor from the variac by a clutch as soon as the signal from the measuring unit ceases. The main disadvantage of this type of regulator is the low life of the contact points of the relays.

2.3.4. *Phase controlled AC voltage regulator*

Voltage regulators using SCRs are quit common. The load voltage is regulated by controlling the firing instants of the SCRs. There are various circuits for single phase and three phase regulators using SCRs. Though the output voltage can be precisely controlled by this method, the harmonic introduced in the load voltage are quit large and this circuit is used for applications where the output voltage waveform need not be strictly sinusoidal. The circuit arrangement for a single phase SCR regulator is shown in Fig. 6 and Fig. 7.

2.3.5. *Ferro-resonant AC voltage regulator*

The concept of the stabilization of AC voltage using a saturated transformer is rather old. The basic circuit arrangement consists of a linear reactor or transformer T1 and a nonlinear saturated reactor or transformer T2 connected in series as shown in Fig. 8. Since the two elements T1 and T2 are in series, the current through them is the same. Transformer T2 is operated under saturated. The voltage division between the two is according to their impedances. Due to nonlinear characteristics of T2 the percentage change of voltage across it is much smaller compared to the percentage change of input voltage. If a suitable voltage proportional to the current is subtracted from the voltage across T2 a practically constant output voltage can be obtained. The circuit arrangement shown is Fig. 8(a) has some drawbacks such as, no load input current is high, and good output voltage stability cab be achieved only at a particular load. Hence some modifications are necessary to improve its performance. The major modification is to place a capacitor across the saturated transformerT2 that is shown in Fig. 8(b).

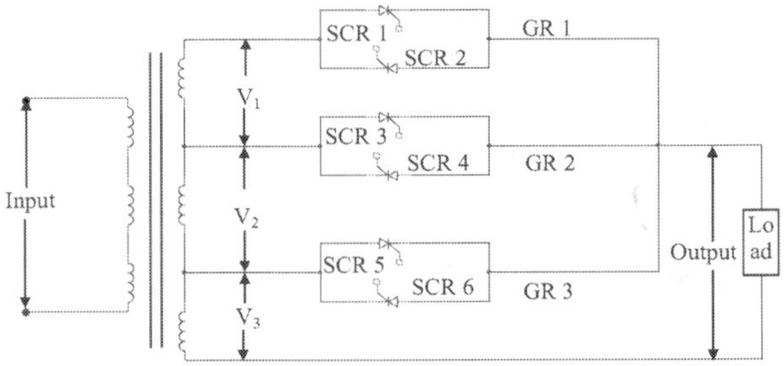

Figure 6. Circuit diagram of solid tap changer using anti-parallel SCRs.

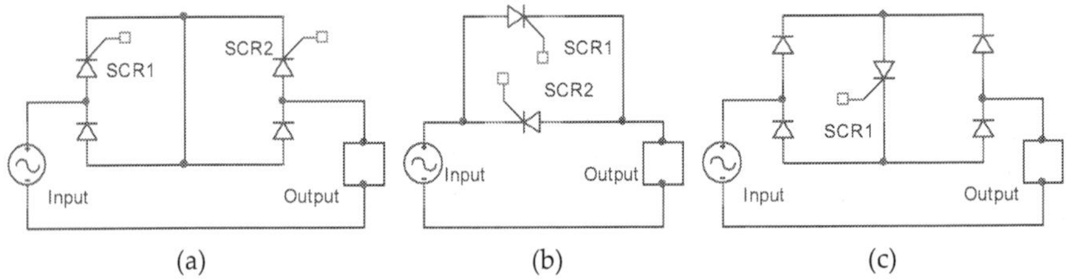

Figure 7. Phase controlled AC voltage regulator, (a) Using back to back SCR and diode and (b) using inverse parallel SCR (c) Using diode-bridge and single SCR

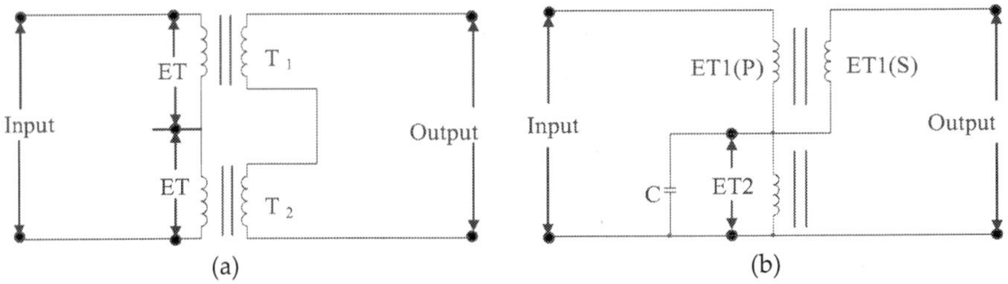

Figure 8. Ferd-resonant AC voltage regulator.

The value of the capacitor is such that it resonant with the saturated inductance of T2 at some point. The characteristics of the circuit is such that a small change in voltage across T2 causes the circuit to go out of resonance consequently a large change in input current and power factor. For a change in the input voltage, the change in voltage across the resonant circuit is small but the change in voltage at T1 is large, and by suitable proportioning of the voltage, a good degree of stabilization is achieved for the variation of input voltage as well as load current. The simple Ferro-resonant regulator has the following disadvantages:

a. The output voltage changes with frequency.
b. Since the core operates in saturation and output is derived from the tank circuit, the core volume is large, the core losses are high and external magnetic field is also high.

2.3.6. Switch mode AC voltage regulator

In switch mode AC voltage regulator, the switching devices are continuously switching on and off with high frequency in order to transfer energy from input to output. The high operating frequency results in the smaller size of the switch mode power supplies since the size of power transformer inductors and filter capacitors is inversely proportional to the frequency. The SMPS are more complicated and more expensive, their switching current can cause noise problems, and simple designs can have a poor power factor.

Four common types of switch mode converters are used in DC-DC conversion. They are Buck, Boost, Buck-Boost and C^UK converters. Researches are trying to modify these DC regulators to regulate AC voltages. Buck- Boost and Cûk converter configuration has been investigated for voltage regulation [17-18]. But in every case it is found that the input power factor is very low and the efficiency is poor.

3. Design and analysis of switching-mode AC voltage regulator

3.1. Operation principle of switching-mode AC voltage regulator

3.1.1. Operation of power circuit

Voltage sag is an important power quality problem, which may affect domestic, industrial and commercial customers. Voltage sags may either decrease or increase in the magnitude of system voltage due to faults or change in loads. Momentary and sustained over voltage and under voltage may cause the equipment to trip out, which is highly undesirable in certain application. In order to maintain the load voltage constant in case of any fluctuation of input voltage or variation of load some regulating device is necessary.

In this chapter the principle of operation of high frequency switching AC voltage regulator, design of its filter circuit and snubber circuit are described. Performance of the regulator is also analyzed using simulation software OrCAD version 9.1. Switch-mode power supplies (SMPS) incorporate power handling electronic components which are continuously switching on and off with high frequency in order to provide the transfer of electric energy from input to output. The design of AC voltage regulator depends on power requirement, degree of stability and efficiency. Solid state AC regulator using phase control technique are not new and are widely used in many application such as heating and lighting control etc.

These regulators are not suited for critical loads because the output waveforms are truncated sine waves, which contain large percentage of distortion. The input power factor is low. These drawbacks are largely overcome and the voltage can be efficiently controlled by means of a solid-state AC regulator using PWM technique.

The power circuit of the proposed AC voltage regulator is shown in Fig. 9. The circuit operation can be explained with the help of Fig. 10. During positive half cycle of the input voltage, at mode 1, when switch-1 is on and switch-2 is off, the current passes through diode D1, switch-1, diode D4 and through the inductor and the energy is stored in the inductor. At mode 2, when switch-1 is off and switch-2 is on, the energy stored in the inductor is transferred through diode D8, switch-2 and diode D5. At mode 1, power is transferred from source and at mode 2, power is not transferred from the source, so by controlling the on and off duration of switch-1 output power can be controlled.

During negative half cycle of the input voltage, at mode 1, when switch-1 is on and switch-2 is off the current passes through the inductor, diode D3, switch-1, and diode D2 and the energy is stored in the inductor. At mode 2, when switch-1 is off and switch-2 is on the energy stored in the inductor is transferred through diode D6, switch-2 and diode D7.

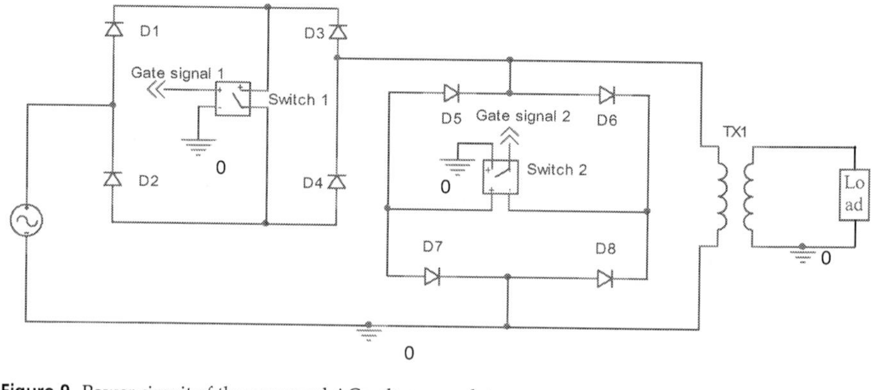

Figure 9. Power circuit of the proposed AC voltage regulator.

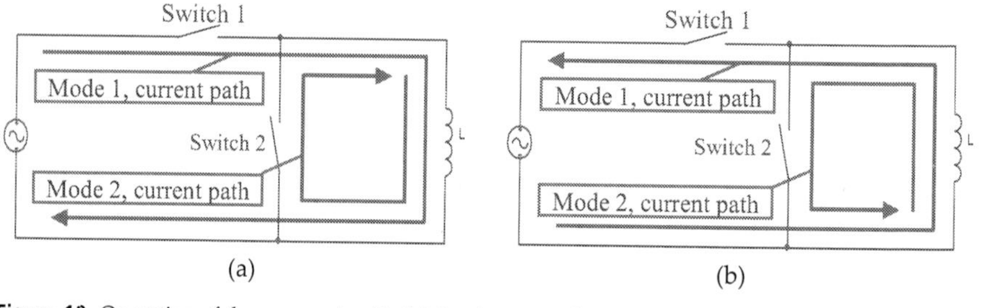

Figure 10. Operation of the power circuit of AC voltage regulator (a) Operation during positive half cycle (b) Operation during negative half cycle

3.1.2. Operation of control circuit

The gate signal generating circuit for a manually controlled AC voltage regulator is shown in Fig. 11. The control circuit incorporates an Operational Amplifier (OPAMP) whose positive input is a variable DC voltage V1, and negative input is a fixed saw-tooth signal V2. In this circuit the OPAMP acts as a comparator, output of the OPAMP depends on the difference of the two inputs. The negative input (saw-tooth wave) is kept constant and positive input (DC voltage) is varied. So output pulse width depends on DC input voltage of OPAMP i.e. when DC input is higher the output of comparator will be wider and when DC input is lower the output of comparator will be narrower.

The outputs of OPAMP are used to turn on/off the switches of the power circuit of the regulator to regulate the output voltage. The output of OPAMP is directly passed through limiter-1 which is the gate signal for switch-1 and after inverting the output of the comparator is passed through the limiter-2 which is the gate signal for switch-2. The function of the limiter is to limit the output of comparator from 0 to 5 V. When switch-1 of the power circuit is on then switch-2 should be off. So the gate signal generating circuit is arranged in such a way that when gate signal of switch-1 is on then gate signal for switch-2 is off and vice versa.

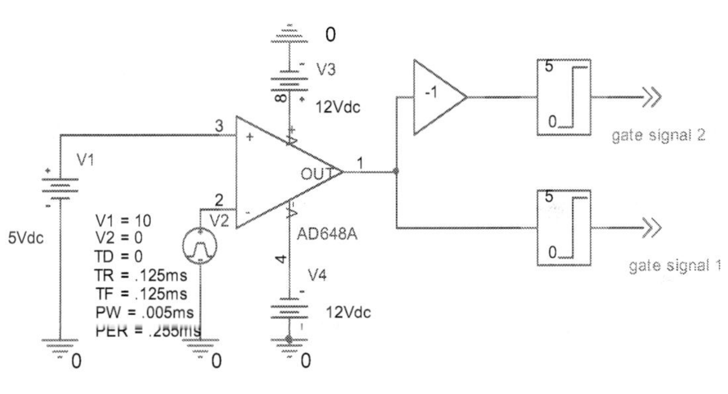

Figure 11. Gate signal generation circuit of manually controlled AC voltage regulator.

3.2. Manually controlled AC voltage regulator

When the stability is not very stringent, manually controlled AC voltage regulator is generally preferred from economic considerations.

The basic circuit of a manually controlled AC voltage regulator is shown in Fig. 12. When any change in output voltage occurs due to change in input voltage or change in load, the voltage can be regulated to the desirable value by changing the DC voltage of the gate signal generating circuit manually. The power circuit of the proposed regulator shown in Fig. 12 is implemented using ideal switches; later part of this chapter the ideal switches is replaced by practical switches. The regulator proposed in this chapter is employed to regulate the output voltage to 300V (peak) for variations of input voltage from 200V (peak) to 400V (peak), also for variation of load from 100 ohm to 200 ohm. However, the output voltage can be set to any desirable value according to requirement. The values of all voltages and currents indicated in this chapter are in peak values.

The input current and output voltage waveforms of the manually controlled AC voltage regulator as shown in Fig. 12, is shown in Fig. 13, when the input voltage is 300V and output voltage is also 300V. The spectrum of the input current and output voltage as shown in Fig. 13 is shown in Fig. From the waveforms it is seen that the waveforms are not smooth, sinusoidal and from the spectrum it is seen that due to high frequency switching the significant number and amount of harmonics occur. The switching frequency is selected to 4 KHz. Harmonics occurs at switching frequency and its odd multiple frequencies. So, filters are required at input and output side to filter out these harmonics to get desired sinusoidal waveforms.

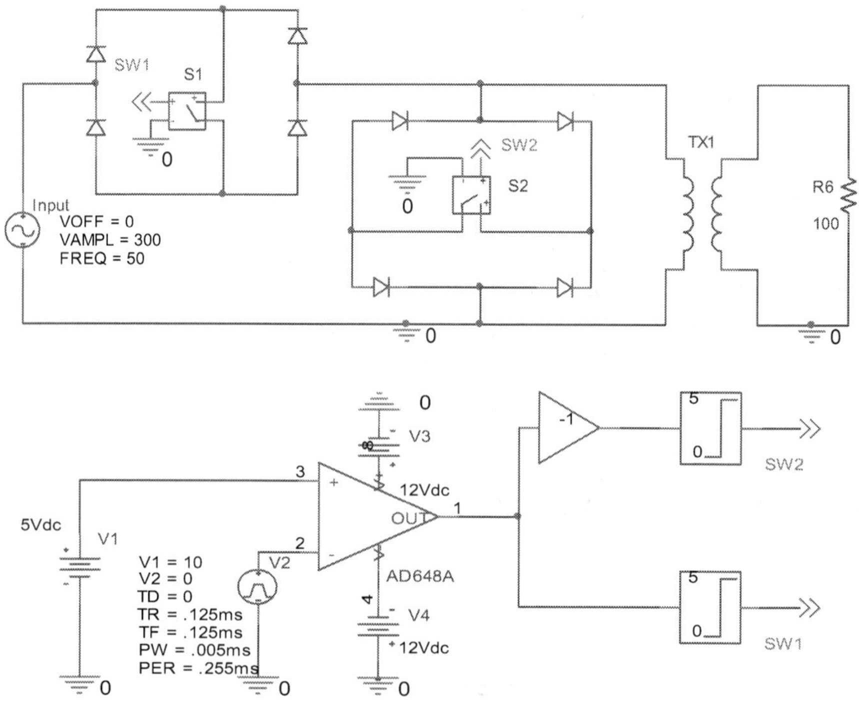

Figure 12. Fundamental circuit of manually controlled AC voltage regulator (ideal switch implementation).

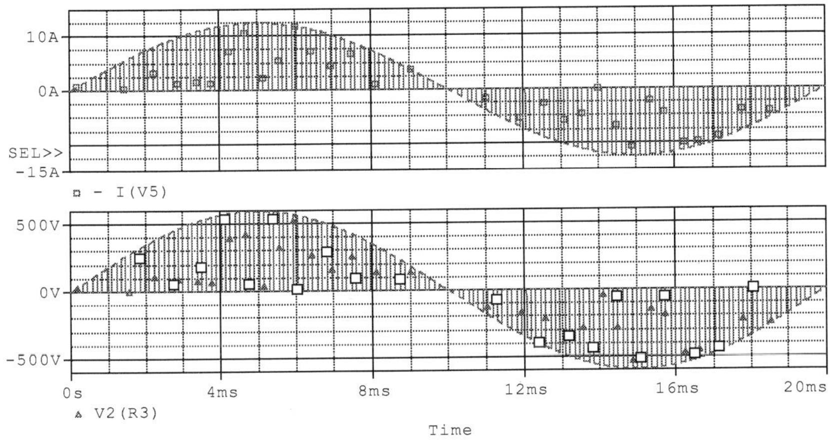

Figure 13. Input current and output voltage waveforms of the regulator shown in Fig. 12. -I(V5): Input current, V(R3:2): Output voltage.

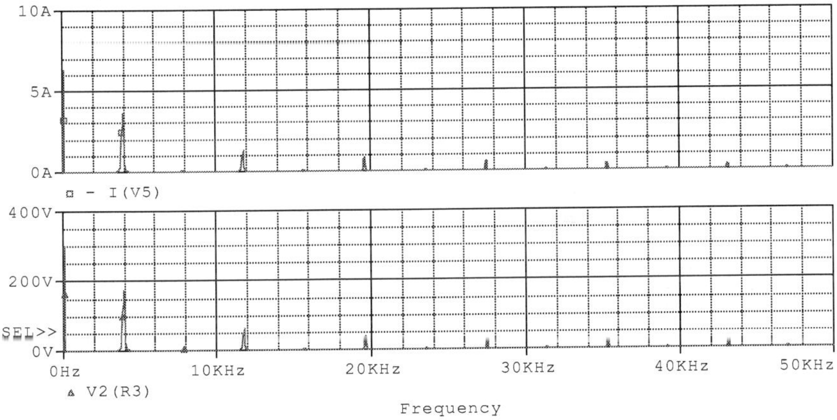

Figure 14. Spectrum of input current and output voltage waveforms. -I(V5):Input current V(R3:2): Output voltage.

3.3. Filter design

3.3.1. Output filter design

For getting smooth output voltage, a low pass LC filter of proper L and C value is needed at the output of this regulator. The output filter circuit and the corresponding AC sweep are shown in Fig. 15. From this circuit we can write, $V_0 = \dfrac{V_{in} \times J/\omega C}{J(\omega L - 1/\omega C)}$ or $\dfrac{V_0}{V_{in}} = \dfrac{1}{\omega C(\omega L - 1/\omega C)}$

The input to the filter is high frequency modulated 50 Hz AC input. The switching signal that modulates the 50 Hz signal is taken to be 4 KHz in this case. So, we will have to make a filter that would pass signal up to 1 KHz (say) and attenuate all other frequencies. This would result a nearly sinusoidal output voltage. In the LC filter section we choose a capacitor of 5μF and determine the value of inductor for a cutoff from AC sweep analysis through OrCAD simulation. We found the value of the inductor to be 30 mH.

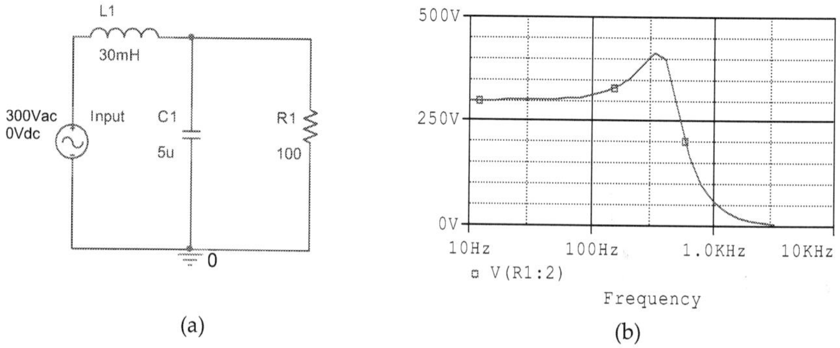

Figure 15. Output voltage filter and AC sweep analysis (a) Output voltage filter (b) AC sweep analysis.

3.3.2. *Input filter design*

A low pass LC filter of proper L and C value is needed at the input of the regulator to filter out some of the harmonics from the supply system. The input filter circuit and the corresponding AC sweep are shown in Fig. 16. Input current contains harmonics at switching frequency 4 KHz and its odd multiple. In order to remove harmonics above 1 KHz, we choose a capacitor of 5μF and determine the value of inductor for a cutoff from AC sweep analysis through OrCAD simulation. We found the value of the inductor to be 30 mH.

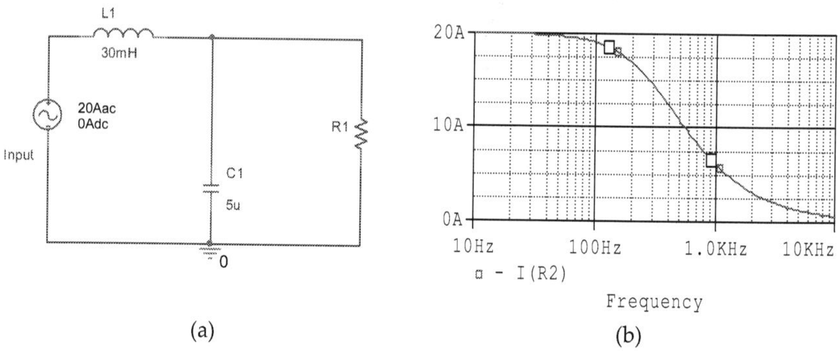

Figure 16. Input current filter and corresponding AC sweep analysis (a) Input current filter (b) AC sweep analysis.

3.4. Free wheeling path and surge voltage across switching devices

The power circuit of the proposed regulator with input and output filter is shown in Fig. 17. In an inductor, current does not change instantaneously. When the switches of power circuit switched on and off the current into the inductor of input and output filter are changed abruptly. Abrupt change of current causes a high di/dt resulting high voltage which is equal to Ldi/dt. These voltages appear across the switches as surge. Usually providing freewheeling path in restricts such occurrence.

3.4.1. Surge voltage across switches

In the proposed circuit, two switches serve as the freewheeling path for each other. However, for very short period when one switch is turned off and other is turned on, an interval elapses due to delay in the switching time. As a result, freewheeling during this interval is disrupted in the proposed circuit. If the current in any inductive circuit is abruptly disrupted, a high Ldi/dt across the switch appears due to the absence of freewheeling path. High spiky surge voltage appears across the switches during these short intervals as shown in Fig. 18.

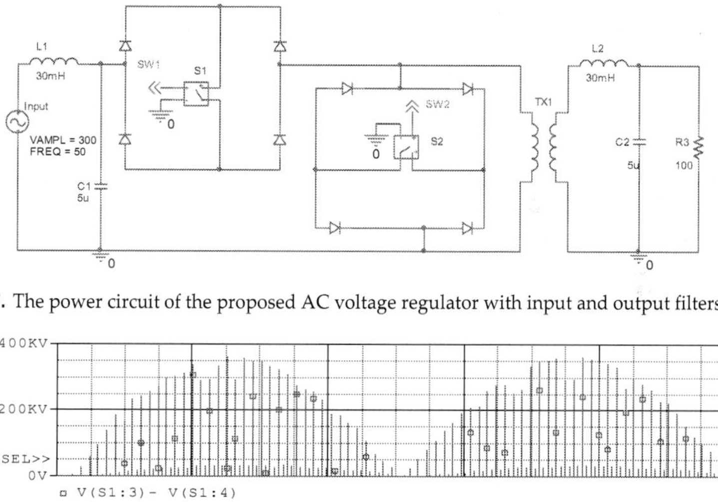

Figure 17. The power circuit of the proposed AC voltage regulator with input and output filters.

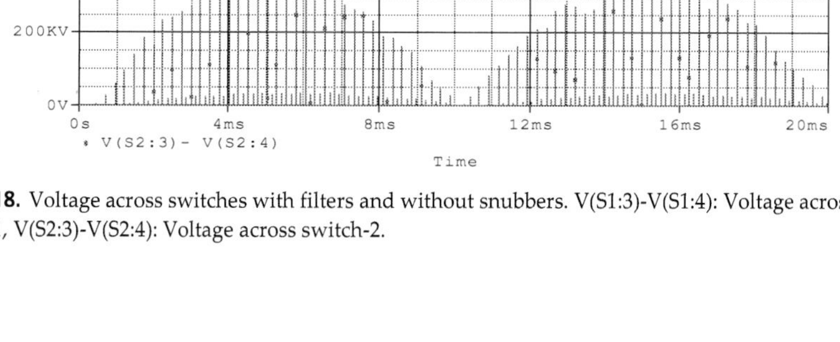

Figure 18. Voltage across switches with filters and without snubbers. V(S1:3)-V(S1:4): Voltage across switch-1, V(S2:3)-V(S2:4): Voltage across switch-2.

These spiky voltages across the switches may be excessively high, about thousand of kilovolt and which may destroy the switches during the operation of the circuit. Remedial measures should be taken to prevent this phenomenon to make the circuit commercially viable. In the proposed circuit RC snubbers are used for suppressing surge voltage across the switches. The power circuit of the proposed regulator with input output filter and snubbers is shown in Fig. 19.

Snubber enhances the performance of the switching circuits and results in higher reliability, higher efficiency, higher switching frequency, smaller size and lower EMI. The basic intent of a snubber is to absorb energy from the reactive elements in the circuit. The benefits of this may include circuit damping, controlling the rate of change of voltage or current or clamping voltage overshoot. The waveforms of voltages across switches with input output filters and snubbers are shown in Fig. 20.

Use of snubbers reduces the spiky voltage across the switches to a tolerate limit for practical application of the AC voltage regulator.

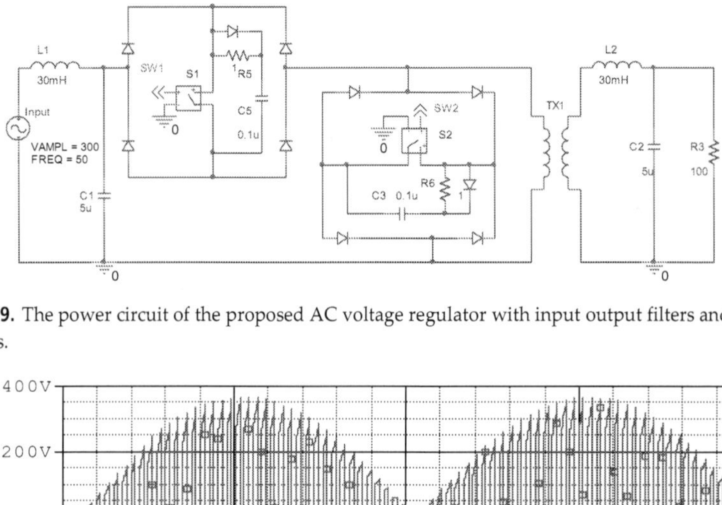

Figure 19. The power circuit of the proposed AC voltage regulator with input output filters and snubbers.

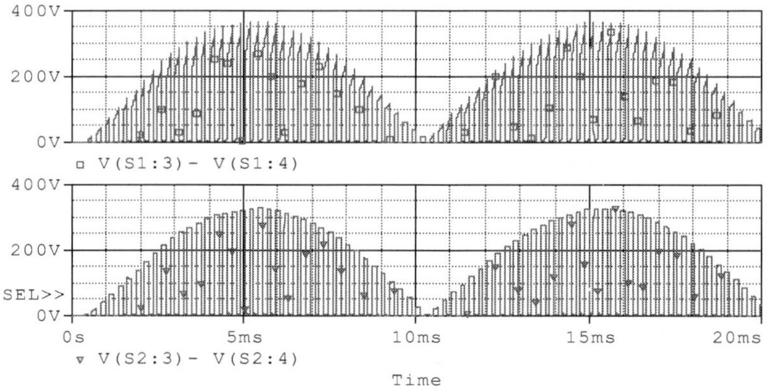

Figure 20. Voltage across switches with filters and snubbers. V(S1:3)-V(S1:4): Voltage across switch-1, V(S2:3)-V(S2:4): Voltage across switch-2.

3.5. Proposed AC voltage regulator with practical switches

In the previous section, we have studied the regulator using ideal S-break switches which have been operated by the pulses from the limiter. But for practical application, real switches are essential which are to be controlled by the pulses having ground isolation. The proposed AC voltage regulator circuit with practical switches is shown in Fig. 21. The ideal S-break switch is replaced by IGBT.

In the proposed regulator chip SG1524B is used to control the gate signal. Signal from the chip is fed to the Limiter and finally to the optocoupler. The output of the optocoupler is used to control the on off time of the IGBTs. The function of the Limiter is to limit the output voltage of the gate signal generating IC from 0 to 6 volts. Optocoupler is used to generate signaling voltage with ground isolation.

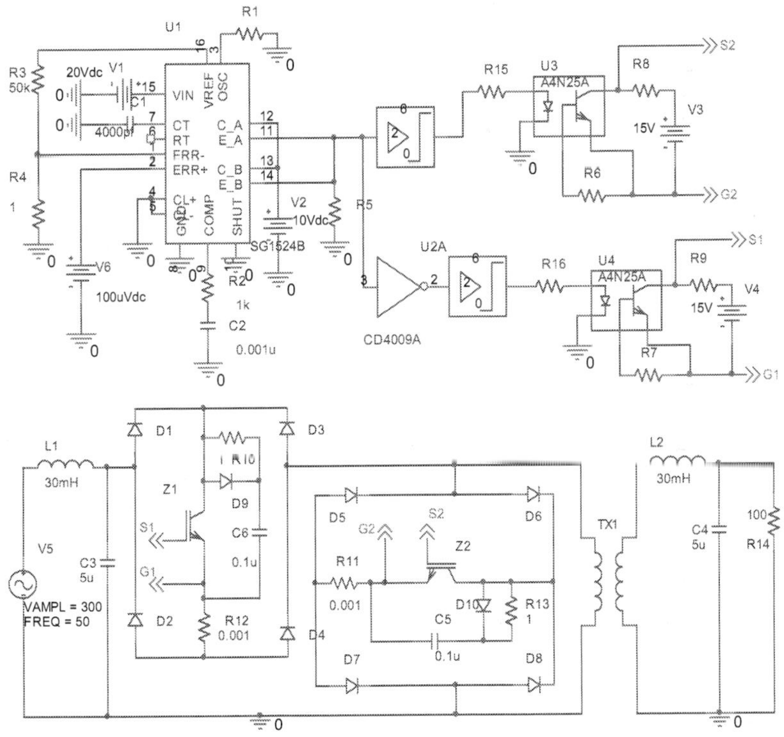

Figure 21. Manually controlled AC voltage regulator circuit with practical switches.

3.5.1. Chip SG1524B for generation of gate signal

The Block diagram of the internal circuitry of the chip SG1524B is shown in Fig. 22. By controlling the error signals of the error amplifier the duty cycle of the gate signal to the regulator can be controlled. Thus it is a very suitable device for using in the regulator circuits.

3.5.2. Results of proposed AC voltage regulator (practical switch implementation)

The waveforms of the input and output voltages of the proposed regulator are shown in Fig. 23 and Fig. 24. Fig. 23 shows the input and output voltages waveform when the input voltage is 200V and output voltage is 300V. Fig. 24 shows the input and output voltages waveform when the input voltage is 400V and output voltage is 300V. Fig. 25 and Fig. 26 show the input and output current waveforms corresponding to Fig. 23 and Fig. 24.

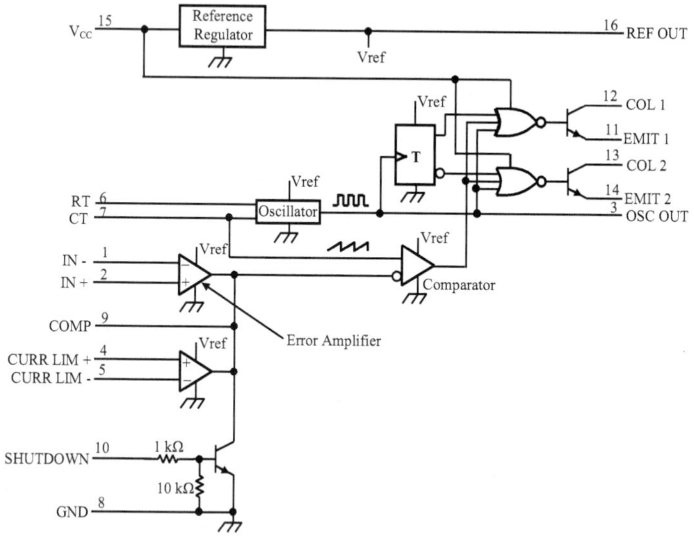

Figure 22. Block diagrm of IC chip SG1524B

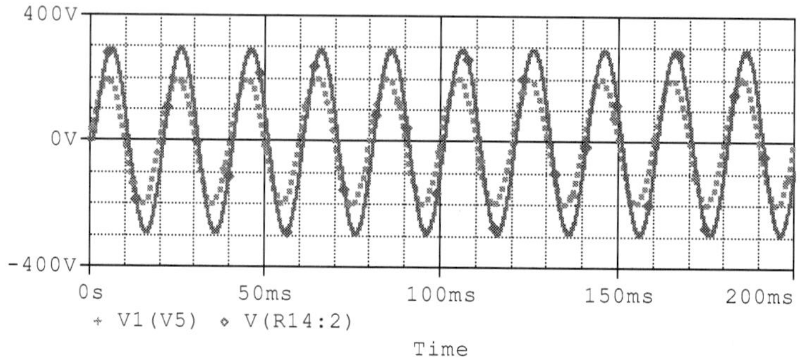

Figure 23. Input and output voltage waveforms, Input 200V output 300V. V1(V5): Input voltage – dotted line, V(R14:2): Output voltage – solid line.

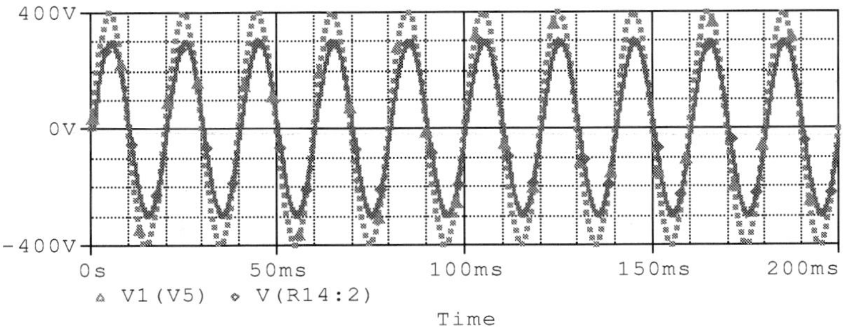

Figure 24. Input and output voltage waveforms, Input 400V output 300V. V1(V5): Input voltage – dotted line, V(R14:2): Output voltage – solid line.

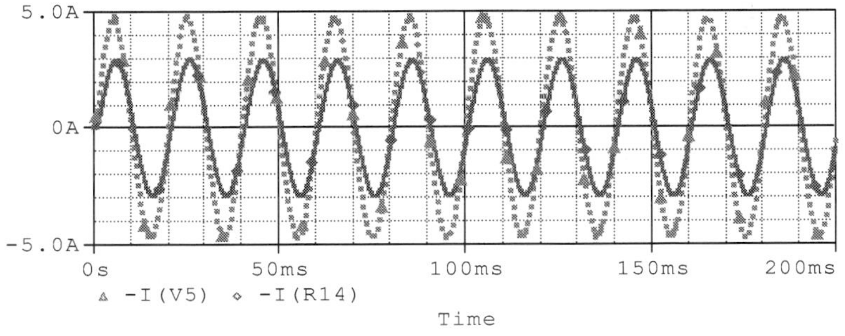

Figure 25. Input and output current waveforms for input 200V output 300V. -I(V5): Input current - dotted line, -I(R14): Output current – solid line.

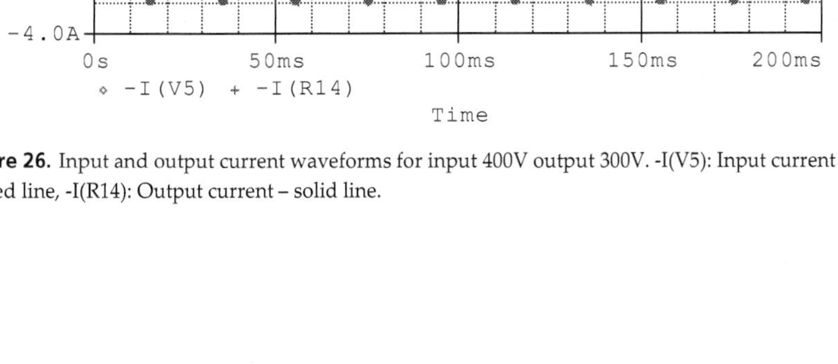

Figure 26. Input and output current waveforms for input 400V output 300V. -I(V5): Input current – dotted line, -I(R14): Output current – solid line.

From the waveforms shown in Fig. 23 to Fig. 26, it is seen that the waveforms of output voltage and input current is perfectly sinusoidal. The variation of output voltage of the proposed regulator with the duty cycle is shown in Fig. 27. The value of input voltage is kept constant to 300V. From Fig. 23 it is seen that the variation of output voltage with duty cycle is almost linear. The variation of duty cycle with the variation of input voltage from 200V to 400V to maintain the output voltage constant to 300V is shown in Fig. 28.

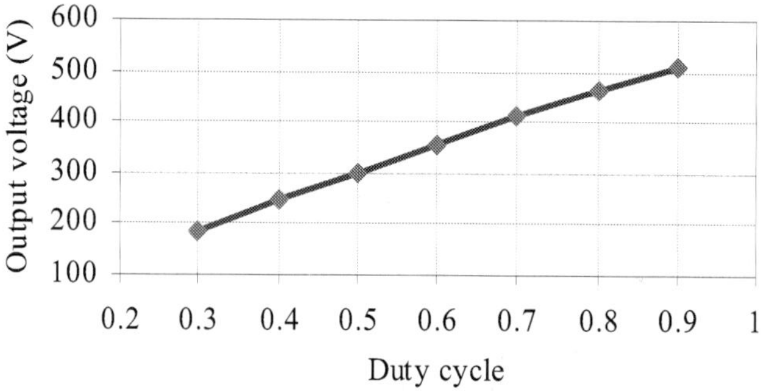

Figure 27. Variation of output voltage with duty cycle. Input voltage is 300V.

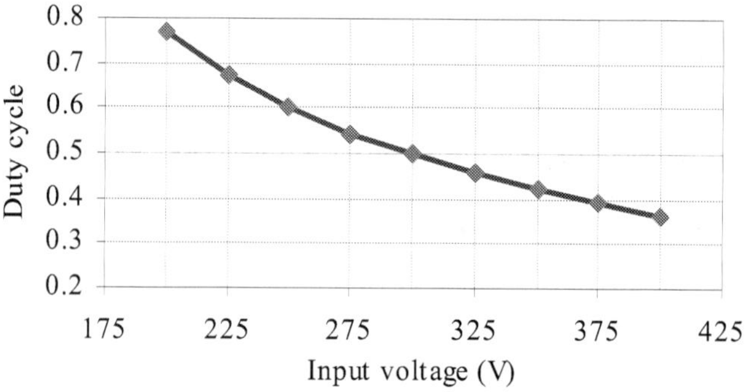

Figure 28. Variation of duty cycle with input voltage to maintain output voltage constant to 300V.

4. Automatic controlled AC voltage regulator

In manually controlled AC voltage regulator control, the output voltage is sensed with a voltmeter connected at the output; the decision and correcting operation is made by a human judgment. The manual control may not be feasible always due to various factors. In automatic voltage regulators, all functions are performed by instruction, and give much better performance, so far as stability, speed of correction, consistency, fatigue, etc. are concerned.

There are two types of automatic control voltage regulator, discontinuous control and continuous control. The automatic control system consists of a sensing or measuring unit and a power control or regulating unit. The sensing unit compares the output voltage or the controlled variable with a steady reference and gives an output proportional to their difference called the error signal. The error voltage is amplified, integrated or differentiated or modified whenever necessary. The processed error voltage is fed to the main control unit to have required corrective action.

In the discontinuous type of control, the measuring unit is such as to produce no signal as long as the voltage is within certain limits. When the voltage goes outside this limit, a signal is produced by the measuring unit until the voltage is again brought within this limit. In this type of measuring or sensing unit, the correcting voltage is independent of percentage of error. When the voltage is brought back to this limit, the signal from the measuring unit is zero and the regulating unit remains at its new position until another signal is received from the measuring unit.

In continuous control, the measuring unit produces a signal with amplitude proportional to the difference between the fixed reference and the controlled voltage. The output of the measuring unit is zero when the controlled voltage or a fraction of it is equal to the reference voltage. The regulating or the controlling unit, which is associated with the continuous measuring unit, gives a correcting voltage proportional to the output of the measuring unit. The principle of operation of a continuous control AC voltage regulator is described in this section.

4.1. Control and gate signal generating circuit for controlled AC voltage regulator

Figure 29 shows the circuit of the proposed automatic controlled AC voltage regulator including the control and gate signal generating circuit. A fraction of the output voltage after capacitor voltage dividing and rectifying is passed through an OPAMP buffer. Buffer is used to remove the loading effect. Output voltage of the buffer is same as its input voltage The output voltage of the buffer is further reduces using resistive voltage divider and taken as the negative input of the error amplifier of the PWM voltage regulating IC SG1524B.

The positive input of the error amplifier is taken from the reference voltage of the chip, after voltage dividing using 50K and 1 ohm resistance. The positive input of the error amplifier is fixed and the negative input is error signal which will vary according to the output voltage. Since the error signal is applied to the negative input of the error amplifier, the duty cycle will be increased if the error signal is decreased and vice versa.

When the output voltage increases above the set value which is 300V either due to change in input voltage or load, the error signal will be increased, therefore the duty cycle will decrease. As a result less power will be transferred from the input to output, and output voltage start to decrease until it reaches to the set value.

When the output voltage decreases below the set value either due to change in input voltage or load then error signal will be decreased which will increase the duty cycle. As a result, more power will be transferred from the input to output, and output voltage start to increase until it reaches to the set value.

When the output voltage is same as the set value than the negative and positive input of the error amplifier will be same as a result the duty cycle will remain same and output voltage will remain unchanged. In this way the proposed regulator will maintain output voltage constant, irrespective of the variation of input voltage and load.

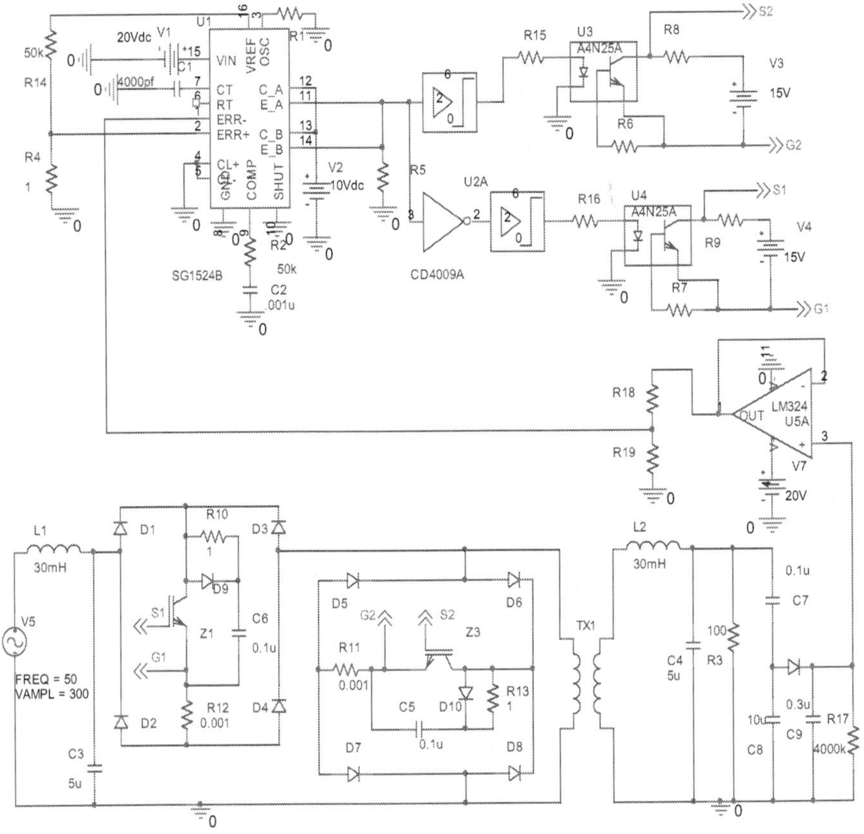

Figure 29. Automatic controlled AC voltage regulator circuit with practical switches.

4.2. Results of automatic controlled AC voltage regulator

Figure 30 shows the input and output voltage waveforms of the proposed automatic controlled AC voltage regulator when the input voltage is 250V and output voltage is 300V. Figure 31 shows the input and output voltage waveforms of the proposed regulator when the input voltage is 350V and output voltage is 300V. Figure 32 and Fig. 33 shows the waveforms of the input current and output currents corresponding to the waveforms of Fig. 30 and Fig. 31 for a load of 100 Ω.

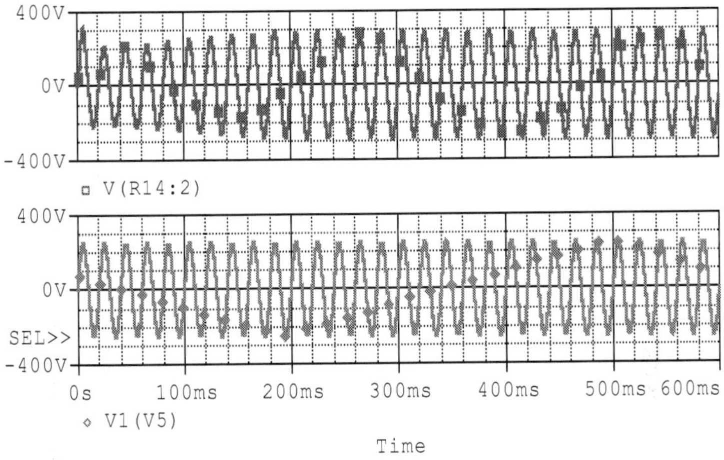

Figure 30. Input and output voltage waveforms for input 250V and output 300V. V1(V5): Input voltage-bottom figure, V(R14:2): Output voltage – top figure.

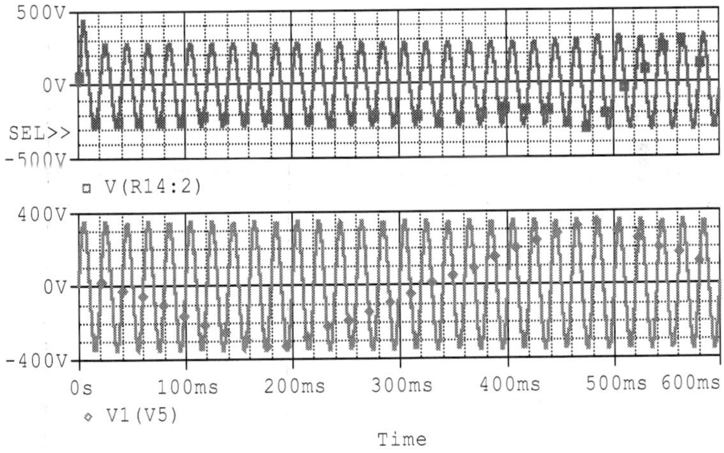

Figure 31. Input and output voltage waveforms for input 350V and output 300V. V1(V5): Input voltage-bottom figure, V(R14:2): Output voltage – top figure.

Table 1 summarizes the result of the proposed regulator to regulate output voltage to 300V for variation of input voltage from 200V to 350V and load from 100 ohm to 200 ohm. In this table input current, output current, input power factor, and efficiency of the regulator are also provided. The proposed regulator can regulate the output voltage effectively, for a wide variation of input voltage and load with efficiency of more than 90% and input power factor more than 0.9.

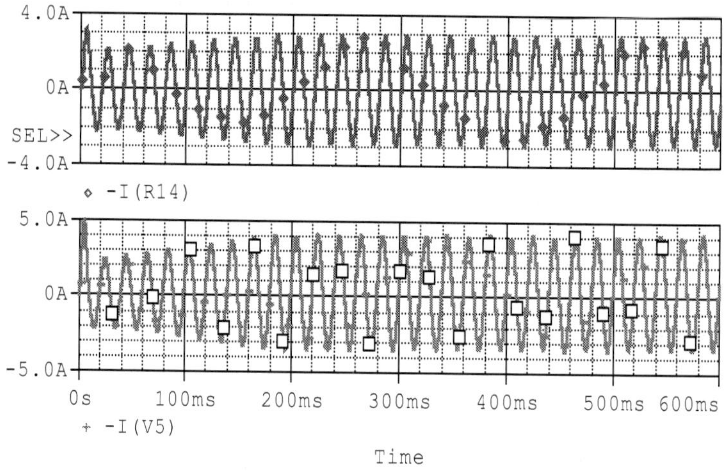

Figure 32. Input and output current waveforms for input 250V output 300V. -I(V5): Input current – bottom figure, -I(R14): Output current – top figure.

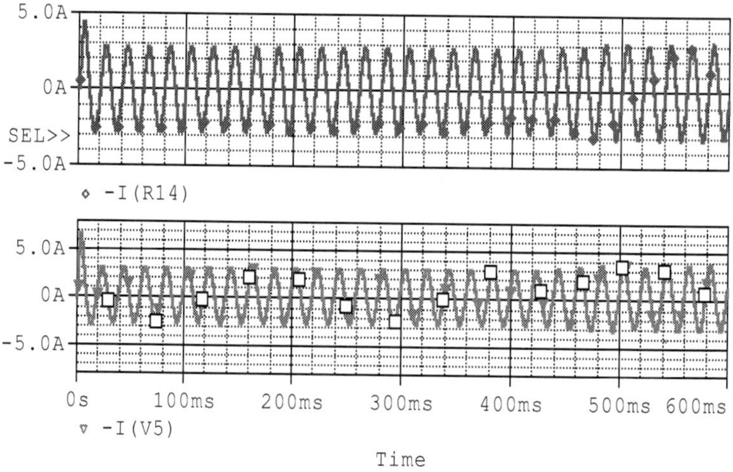

Figure 33. Input and output current waveforms for input 350V output 300V. -I(V5): Input current – bottom figure, -I(R14): Output current – top figure.

V_{in} (V)	I in (A)	Input pf	P_{in} (W)	V_{out} (V)	Load (Ω)	I_{out} (A)	P_{out} (W)	Efficiency (%)
200	4.81	1.00	481	295	100	2.95	435.13	90.46
225	4.30	1.00	483.09	298	100	2.98	444.02	91.91
250	3.92	1.00	489.70	300	100	3.00	450.00	91.89
275	3.60	1.00	495.00	300	100	3.00	450.00	90.91
300	3.30	1.00	493.79	300	100	3.00	450.00	91.13
325	3.10	0.99	498.85	302	100	3.02	456.02	91.41
350	2.95	0.98	508.41	305	100	3.05	465.13	91.49
250	2.07	0.96	248.73	300	200	1.50	225.00	90.46
275	1.90	0.95	248.46	300	200	1.50	225.00	90.56
300	1.75	0.95	248.20	300	200	1.50	225.00	90.65
325	1.68	0.93	253.12	302	200	1.51	228.01	90.08
350	1.60	0.91	253.77	305	200	1.53	232.56	91.64

*All voltages and currents values in this table are in peak values.

Table 1. Results of proposed automatic controlled AC voltage regulator for maintaining output 300 V.

5. Conclusion

An essential feature of efficient electronic power processing is the use of semiconductors devices in switch mode to control the transfer of energy from source to load through the use of pulse width modulation techniques. Inductive and capacitive energy storage elements are used to smooth the flow of energy while keeping losses at a lower level. As the frequency of the switching increases, the size of the capacitive and inductive elements decreases in a direct proportion. Because of the superior performance, the SMPS are replacing conventional linear power supplies.

In this chapter the design and analysis of an AC voltage regulator operated in switch mode is described in details. AC voltage regulator is used to maintain output voltage constant either for an input voltage variation or load variation to improve the power quality. If the output voltage remains constant, equipment life time increases and outages and maintenance are reduced.

At first the regulator is analyzed using ideal switches, then the ideal switches is replaced by practical switches which required isolated gate signal. The procedure of smoothing the input current and output voltage, and suppressing the surge voltage across the switches is described. A manually controlled AC voltage regulator is analyzed then the concept of operation of an automatic controlled AC voltage regulator is described. Finally an automatic controlled AC voltage regulator is designed and its performance is analyzed.

The proposed regulator can maintain the output voltage constant to 300V, when input voltage is vary from 200V to 350V also for variation of load. To maintain constant output voltage PWM control is used. By varying the duty cycle of the control circuit have achieved the goal of maintaining the constant output voltage across load. For generation of gate signal of the switches an IC chip SG1524B is used which is compact and commercially available at a very low cost. The input current of the proposed regulator is sinusoidal and the input power factor is above 0.9. From simulation results it is seen that the efficiency of the proposed regulator is more than 90%.

Author details

Raju Ahmed
Electrical and Electronic Engineering Department, Dhaka University of Engineering and Technology (DUET), Gazipur, Bangladesh

Mohammad Jahangir Alam
Electrical and Electronic Engineering Department, Bangladesh University of Engineering and Technology (BUET), Dhaka, Bangladesh

6. References

[1] P. B. Steciuk and J. R. Redmon, "Voltage sag analysis peaks customer service," IEEE Comput. Appl. Power, vol. 9, pp. 48-51, Oct. 1996.

[2] M. F. McGranaghan, D. R. Mueller, and M. J. Samotyj, "Voltage sags in industrial systems," IEEE Transactions on Industrial Application., vol. 29, pp. 397-403, Mar./Apr. 1993.

[3] M. H. J. Bollen, "The influence of motor reacceleration on voltage sags," IEEE Transactions on Industrial Application, vol. 31, pp. 667-674, July/Aug. 1995.

[4] M. H. J. Bollen, "Characterization of voltage sags experienced by three phase adjustable-speed drives," IEEE Transactions on Power Delivery, vol. 12, pp. 1666-1671, Oct. 1997.

[5] H. G. Sarmiento and E. Estrada, "A voltage sag study in an industry with adjustable speed drives",IEEE Industry Applications Magazine., vol. 2, pp. 16-19, Jan./Feb. 1996.

[6] N. Kutkut, R. Schneider, T. Grand, and D. Divan, "AC voltage regulation technologies," Power Quality Assurance, pp. 92-97, July/Aug. 1997.

[7] D. Divan, P. Sulherland, and T. Grant, "Dynamic sag corrector: A new concept in power conditioning," Power Quality Assurance, pp.42-48, Sept./Oct. 1998.

[8] A. Elnadt and Magdy M. A. Salama, "Unified approach for mitigating voltage sag and voltage flicker using the DSTATCOM," IEEE Transactions on Power Delivery, vol. 30, no. 2, April 2005.

[9] S. M. Hietpas and R. Pecan, "Simulation of a three-phase boost converter to compensate for voltage sags, " in Proceeding. IEEE 1998 Rural Electric Power Conference, pp. B4-1-B4-7, Apr. 1998.

[10] S. M. Hietpas and Mark Naden, "Automatic voltage regulator using an AC voltage-voltage converter," IEEE Transactions on Industrial Application, vol. 36, pp. 33-38, Jan.\Feb. 2000.

[11] F. Z. Peng, Lihue Chen, and Fan Zhang, "Simple topologies of PWM AC-AC converters," IEEE Power Electronics Letters, vol. 1, no.1, March 2003.

[12] G. Venkataramanan. B. K. Johnson, and A. Sundaram, "An AC-AC power converter for custom power applications," IEEE Transactions on Power Delivery, vol. 11, pp. 1666-1671, July 1996.

[13] V. Nazquez, A. Velazquez, C. Hernandez, E. Rodríguez and R. Orosco, "A Fast AC Voltage Regulator,", CIEP 2008. 11th IEEE International Power Electronics Congress, pp. 162-166, Aug. 2008.

[14] J. Nan, T. Hou-jun, L. Wei and Y. Peng-sheng, "Analysis and control of Buck-Boost Chopper type AC voltage regulator,", IPEC'09. IEEE 6th International Power Electronics and Motion Control Conference, pp. 1019-1023, May 2009.

[15] N. A. Ahmed, M. Miyatake, H. W. Lee and M. Nakaoka, "A Novel Circuit Topology of Three-Phase Direct AC-AC PWM Voltage Regulator" Industry Applications Conference, 2006. 41st IAS Annual Meeting. Conference Record of the 2006 IEEE, pp. 2076-2081, Oct. 2006.

[16] V. Nazquez, A. Velazquez and C. Hernandez, "AC Voltage Regulator Based on the AC-AC Buck-Boost Converter," ISIE 2007. IEEE International Symposium on Industrial Electronics, pp. 533-537, Jun. 2007.

[17] P. K. Banerjee, "Power line voltage regulation by PWM AC Buck-Boost voltage controller," A M.Sc. thesis, Department of EEE, BUET, July, 2002.

[18] A. Hossain, "AC voltage regulation by Cûk switch mode power supply," A M.Sc. thesis, Department of EEE, BUET, July, 2003.

[19] Li B. H., Choi S.S., Vilathgamuwa D. M., Design considerations on the line-side filter used in the dynamic voltage restorer, IEE Proceedings - Generation, Transmission, and Distribution, pp. 1-7, 2001.

[20] Wang Jing, Xu Aiqin, Shen Yueyue., A Survey on Control Strategies of Dynamic Voltage Restorer, 13th International Conference on Harmonics and Quality of Power (ICHQP), pp. 1-5, Sept. 28 2008-Oct. 1, 2008,.

[21] Nielsen J.G., Blaabjerg F., A Detailed Comparison of System Topologies for Dynamic Voltage Restorers, IEEE Transactions on Industry Applications, vol, 41, no. 5, pp. 1272-1280, 2005.

[22] OrCAD Software, Release 9: 1985-1999 OrCAD, Inc., U.S.A.

[23] Slobodan Cûk, "Basics of Switched Mode Power Conversion Topologies, Magnetics, and Control," Modern Power Electronics: Evaluation, Technology, and applications, Edited by B.K. Bose, IEEE Press, pp. 265-296, 1992.

[24] M. H. Rashid," Power Electronics – Circuits, Devices, and Applications, "Prenctice Hall India, Second Edition, 2000, pp.317-387.

[25] P.C. Sen, "Power Electronics," Tata McGraw-Hill Publishing Company Ltd., India, 1987, pp. 588-614.

[26] R. Thompson, "A Thyristor Alternating Voltage Regulator," IEEE Trans. on Ind. and Gen. Application, vol. IGA 4 (1968) 2, pp. 162-166., 1968.

[27] E. J. Cham and W. R. Roberts, "Current Regulators for Large Rectifier Power Supplies Used on Electrochemical Processing Lines," IEEE Trans. on Ind. and Gen. Application IGA-4 (1968) 6, pp. 609-618.

[28] J. M.(Jr), Mealing , "A Coherent Approach to the Design of Switching Mode DC Regulators," IEEE Conf. Rec. IGA, pp.177-185, Oct. 1967.

[29] Unitrode, "Switching Regulated Power Supply Design Seminar Manual," Unitrode Corporation, U. S. A, 1986.
[30] M. H. Rashid, "A Thyristor Chopper With Minimum Limits on Voltage Control of DC Drives, " International Jurnal of Electronics, Vol. 53, No. 1, pp.71-81, 1982.
[31] K. P. Severns and G. E. Bloom, "Modern DC-to-DC Switch Mode Power Converter Circuits," Van Nostrand Reinholdd Company, Inc., New York, U. S. A, 1983.
[32] S. Cûk, "Survey of Switched Mode Power Supplies," IEEE International Conference on Power Electronics and Variable Speed Drives, London, pp. 83-94, 1985.
[33] M. Ehsani, R. L. Kustom, and R. E. Fuja, "Microprocessor Control of A Current Source DC-DC Converter," IEEE Transactions on Industrial Applications, Vol. LA19, No. 5, pp. 690-698, 1983.
[34] R. D. Middlebrook, "A Continuous Model for the Tapped-Inductor Boost Converter," IEEE Power Electronics Specialists Conference Record, Culver City, CA, U. S. A, pp.63-79, 1975.
[35] Slobodan Cûk R. D. Middlebrook, "A General Unified Approach to Modeling Switching DC-to-DC Converters in Discontinuous Conduction Mode," IEEE Power Electronics Specialists Conference Record, Palo Alto, CA, U. S. A, pp. 36-57, 1977.
[36] R. D. Middlebrook and Slobodan Cûk, "Modeling and Analysis Methods for DC-to-DC Switching Converters," IEEE International Semiconductor Power Converter Conference Record, Lake Buena Vista, FL, U. S. A, pp.90-111, 1977.
[37] Slobodan Cûk and R. D. Middlebrook, "Coupled Inductor and other Extensions of a New Optimum Topology Switching DC-to-DC Converter," IEEE Industry Applications Society Annual Meeting, Record, Los Angeles, CA, U. S. A, pp.1110-1126, 1977.
[38] R. D. Middlebrook and Slobodan Cûk, "Isolation and Multiple Output Extensions of a New Optimum Topology Switching DC-to-DC Converter," IEEE Power Electronics Specialists Conference Record, Syracuse, NY, U. S. A, pp.256-264, 1978.
[39] C. Chen and Deepakraj M. Divan, "Simple Topologies for Single Phase AC Line Conditioning," IEEE Transactions on Industry Applications, Vol.30, No.2, pp. 406-412, March/April 1994.
[40] Mark F. McGranaghan, David R. Mueller and Marek J. Samotyj, "Voltage Sags in Industrial Systems," IEEE Transactions on Industry Applications, Vol.29, No.2, pp. 397-403, March/April 1993.
[41] S. Cûk and R. D. Middlebrook, "Advances in Switched Mode Power Conversion," IEEE Transactions on Industrial Electronics, Vol. IE 30. No. 1, pp.10-29, 1983.
[42] N. Mohan, Tore M. Undeland and William P.Robbins, "Power Electronics- Converters, Applications, and Design, " John Wiley and Sons Inc., Second ed., pp. 161-195 & 669-695, 1995
[43] H. Veffer, "High Current, Low Inductance GTO and IGBT Snubber Capacitors," Siemens Components, June, pp. 81-85, 1990.
[44] B. D. Bedford and R. G. Hoft, "Principles of Inverter Circuits," Wiley: 1964.
[45] E. R. Hnatek, "Design of Solid-State Power Supplies," Van Nostrand Reinhold; 1971.
[46] A. I. Pressman, "Switching and Linear Power Supply," Power Converter Design, vol. I and vol. II, Hayden; 1977.